Contractor and Client Relations to Assure Process Safety

Publications Available from the
CENTER FOR CHEMICAL PROCESS SAFETY
of the
AMERICAN INSTITUTE OF CHEMICAL ENGINEERS

Inherently Safer Chemical Processes: A Life Cycle Approach
Contractor and Client Relations to Assure Process Safety
Guidelines for Use of Vapor Cloud Dispersion Models, Second Edition
Guidelines for Evaluating Process Plant Buildings for External Explosions and Fires
Guidellines for Writing Effective Operating and Maintenance procedures
Guidelines for Chemical Transportation Risk Analysis
Guidelines for Safe Storage and Handling of Reactive Materials
Guidelines for Technical Planning for On-Site Emergencies
Guidelines for Process Safety Documentation
Guidelines for Safe Process Operations and Maintenance
Guidelines for Process Safety Fundamentals in General Plant Operations
Guidelines for Chemical Reactivity Evaluation and Application to Process Design
Tools for Making Acute Risk Decisions with Chemical Process Safety Applications
Guidelines for Preventing Human Error in Process Safety
Guidelines for Evaluating the Characteristics of Vapor Cloud Explosions, Flash Fires, and BLEVEs
Guidelines for Implementing Process Safety Management Systems
Guidelines for Safe Automation of Chemical Processes
Guidelines for Engineering Design for Process Safety
Guidelines for Auditing Process Safety Management Systems
Guidelines for Investigating Chemical Process Incidents
Guidelines for Hazard Evaluation Procedures, Second Edition with Worked Examples
Plant Guidelines for Technical Management of Chemical Process Safety, Rev. Ed.
Guidelines for Technical Management of Chemical Process Safety
Guidelines for Chemical Process Quantitative Risk Analysis
Guidelines for Process Equipment Reliability Data, with Data Tables
Guidelines for Vapor Release Mitigation
Guidelines for Safe Storage and Handling of High Toxic Hazard Materials
Understanding Atmospheric Dispersion of Accidental Releases
Expert Systems in Process Safety
Concentration Fluctuations and Averaging Time in Vapor Clouds
Safety, Health, and Loss Prevention in Chemical Processes: Problems for Undergraduate Engineering Curricula
Safety, Health, and Loss Prevention in Chemical Processes: Problems for Undergraduate Engineering Curricula—Instructor's Guide
Proceedings of the International Conference and Workshop on Modeling and Mitigating the Consequences of Accidental Releases of Hazardous Materials, 1995.
Proceedings of the International Symposium and Workshop on Safe Chemical Process Automation, 1994
Proceedings of the International Process Safety Management Conference and Workshop, 1993
Proceedings of the International Conference on Hazard Identification and Risk Analysis, Human Factors, and Human Reliability in Process Safety, 1992
Proceedings of the International Conference/Workshop on Modeling and Mitigating the Consequences of Accidental Releases of Hazardous Materials, 1991.
Proceedings of the International Symposium on Runaway Reactions, 1989
CCPS/AIChE Directory of Chemical Process Safety Services

Contractor and Client Relations to Assure Process Safety

A CCPS CONCEPT BOOK

William F. Early, II

Early Consulting, L.C.
Houston, Texas

CENTER FOR CHEMICAL PROCESS SAFETY
of the
AMERICAN INSTITUTE OF CHEMICAL ENGINEERS

345 East 47th Street, New York, New York 10017

Library of Congress Cataloging-in Publication Data
Contractor and Client Relations to Assure Process Safety
p. cm.
Includes bibliographical references and index.
ISBN 978-0-8169-0667-3
1. Chemical processes—Safety measures. 2. Contractors operations. 3. Industrial safety. I. American Institute of Chemical Engineers. Center for Chemical Process Safety.
TP149.C66 1996
660' .068'4—dc20 94–2481

This book is available at a special discount when ordered in bulk quantities. For information, contact the Center for Chemical Process Safety of the American Institute of Chemical Engineers at the address shown above.

CONTENTS

PREFACE

The Center for Chemical Process Safety (CCPS) was established in 1985 by the American Institute of Chemical Engineers (AIChE) for the express purpose of assisting the Chemical and Hydrocarbon Processing Industries in avoiding or mitigating catastrophic chemical accidents. To achieve this goal, CCPS has focused its work on four areas:

- establishing and publishing the latest scientific and engineering practices (not standards) for prevention and mitigation of incidents involving toxic and/or reactive materials;
- encouraging the use of such information by dissemination through publications, seminars, symposia and continuing education programs for engineers;
- advancing the state-of-the-art in engineering practices and technical management through research in prevention and mitigation of catastrophic events; and
- developing and encouraging the use of undergraduate education curricula which will improve the safety knowledge and consciousness of engineers.

The current book, *Contractor and Client Relations to Assure Process Safety*, is intended to identify issues between the Contractor and his Client which should be addressed; particularly those which may affect process safety. These issues may arise at any stage during the life of a process facility. This book is concerned with resolving those issues early in the contractor–client relationship so that there can be a clearer project focus on process safety issues. Improved communications and understanding of risks can benefit each area of process safety. The Contractor–Client Relations subcommittee and its contractor, Early Consulting, L.C., have attempted to provide a guide to the literature to assist the reader who wishes to go beyond the process safety issues and address project considerations.

This book has been organized so as to identify and address basic safety programs first. The book then includes a discussion of the engineering–procurement–construction (EPC) contractual bases and split of work as they address OSHA PSM issues, followed by subcontract considerations. Finally, the book presents a treatise on managing contractor–client risk.

ACKNOWLEDGMENTS

The American Institute of Chemical Engineers (AIChE) wishes to thank the Center for Chemical Process Safety (CCPS) and those involved in its operation, including its many sponsors whose funding made this project possible, the members of its Technical Steering Committee who conceived of and supported this *Concept* project, and the members of the Contractor-Client Relations Subcommittee.

The members of the Contracto—Client Relations Subcommittee were Stanley E. Anderson, Rohm & Haas Texas, Incorporated; Paul R. Chaney, Mobil Chemical Company; Lawrence T. Denk, Stone & Webster Engineering Corporation; Al J. McCarthy, The M. W. Kellogg Co., Subcommittee Chairman; Ray E. Witter, CCPS Staff.

The author of this *Concept* book was William F. (Skip) Early, II, P.E., of Early Consulting, L.C.

Book peer reviewers were Janet E. Ritz, A. Lee McLain, P. C.; Keith Thayer and Jerry Power, CDI-Stubbs Overbeck; William W. Doerr, PhD, Dan Smith, and Don Turner, CH2M HILL; L. E. Schmaltz, Exxon Chemical Americas; David F. Montague, JBF Associates, Inc.; Peter N. Lodal, Tennessee Eastman; and William J. Minges, CCPS.

Lastly, the Contractor-Client Relations Subcommittee wishes to express its appreciation to Lester H. Wittenberg and Dr. Jack Weaver of the CCPS for their enthusiastic support.

GLOSSARY

Administrative Controls: Procedural mechanisms, such as lockout/tagout procedures, for directing and/or checking human performance on plant tasks.

Arbitration: The settlement of a dispute by an arbitrator, named by consent of both sides, to listen to the opposing arguments and come to a binding decision.

Catastrophic Incident: An incident involving a major uncontrolled emission, fire or explosion with an outcome effect zone that may impact employees and/or the surrounding community.

Consequential Damages: Damages incurred as an indirect result of an action.

Constructor: A contractor who performs only construction activities on a project. In this case, engineering design may be performed by the client company or another contractor.

Consultant: Typically a small contractor who may provide services, typically professional advice or opinions, related to any aspect of the plant life cycle.

Contract Employee: An employee who is employed directly by the client on a contract basis. This type of employee is typically a self-employed independent contractor who is not covered by company benefits programs, but in other aspects such as work hours or duties may appear to be a direct employee.

Discovery: Discovery is the pre-trial procedure of taking depositions or other means to compel disclosure of pertinent factual information.

EPC Contractor: A contractor who is able to offer the full range or any portion of engineering design, procurement, and construction activities on a single project utilizing in-company resources.

Failure: An unacceptable difference between expected and observed performance.

Fixed Price: A fixed price contract is similar in nature to a lump sum contract, but has a payment schedule which provides a periodic cash flow to the contractor.

Fixed Rate, Reimbursable (Open Cost): Fixed rate, reimbursable denotes that contract employees or contractors work at fixed hourly or daily rates, based upon a stated rate schedule, and expenses are reimbursed at agreed rates or cost.

Hazard: An inherent chemical or physical characteristic that has the potential for causing damage to people, property, or the environment. In this document it is typically the combination of a hazardous material, an operating environment, and certain unplanned events that could result in an accident.

Hazard Analysis: The identification of undesired events that lead to the materialization of a hazard, the analysis of the mechanisms by which these undesired events could occur and usually the estimation of the consequences.

Human Factors: A discipline concerned with designing machines, operations, and work environments so that they match human capabilities, limitations, and needs. Includes any technical work (engineering, procedure writing, worker training, worker selection, etc.) related to the human factor in operator–machine systems.

Indemnity: A protection or insurance against loss, damage, or liability.

Inherently Safe: A system is inherently safe if it remains in a non-hazardous situation after the occurrence of unacceptable deviations from normal operating conditions.

Likelihood: A measure of the expected frequency with which an event occurs. This may be expressed as a frequency (e.g., events per year), a probability of occurrence during a time interval (e.g., annual probability), or a conditional probability (e.g., probability of occurrence, given that a precursor event has occurred).

Lump Sum: A lump sum project is one for which a single, lump sum payment is made. This type of project is usually of short duration due to the time value of money, but it is not limited in time.

Mediation: The process of diplomatically intervening, with the consent of both parties, to assist in bringing a dispute to a resolution.

Mitigation: Lessening the risk of an accident event sequence by acting on the source in a preventive way by reducing the likelihood of occurrence of the event, or in a protective way by reducing the magnitude of the event and/or the exposure of local persons or property.

Process Hazard Analysis: A thorough, orderly, qualitative approach for identifying, evaluating, and controlling the potential hazards of processes.

Process Safety: A discipline that focuses on the prevention of fires, explosions, and accidental chemical releases at chemical process facilities. Excludes classic worker health and safety issues involving working surfaces, ladders, protective equipment, etc.

Process Safety Management (PSM): A program or activity involving the application of management principles and analytical techniques to ensure the safety of process facilities. This is typified by the program in 29 CFR 1910.119 Process Safety Management of Highly Hazardous Chemicals.

Risk: The combination of the expected frequency and consequence of a single accident or group of accidents.

Risk Management: The systematic application of management policies, procedures, and practices to the tasks of analyzing, assessing, and controlling risk in order to protect employees, the general public, the environment, and company assets. In following chapters, the term risk management will be used to apply to both process risks and business risks.

Subcontractor: A contractor who takes on some or all of the obligations of the primary contractor. Examples of subcontractors may include insulators, electricians, plumbers, or specialized consultants.

Time and Materials: Time and materials is similar to fixed rate, reimbursable, in that the contractor's time or manhours are paid by the client at an agreed rate; whether at a multiplier of the contract employee's salary, or at a fixed rate. Other contractor expenses are typically reimbursed at cost or at an agreed markup.

ACRONYMS AND ABBREVIATIONS

AGC	Associated General Contractors (of America)
AIChE	American Institute of Chemical Engineers
API	American Petroleum Institute
ASME	American Society of Mechanical Engineers
CAA	Clean Air Act
CAAA	Clean Air Act Amendments
CCPS	Center for Chemical Process Safety
CFR	Code of Federal Regulations
CII	Construction Industry Institute
CMA	Chemical Manufacturers Association
DOT	Department of Transportation
EPA	Environmental Protection Agency
EPC	Engineering, Procurement, Construction (Contractor)
ISO	International Standards Organization
MOC	Management of Change
MSDS	Material Safety Data Sheet
NASA	National Aeronautics and Space Administration
NDE	Non-destructive Examination
NDT	Non-destructive Testing

NFPA	National Fire Protection Association
OSHA	Occupational Safety and Health Administration
PFD	Process Flow Diagram
P&ID	Piping and Instrumentation Diagram
PHA	Process Hazard Analysis
PMI	Project Management Institute
PSM	Process Safety Management
PSSR	Pre-startup Safety Review
RP	Recommended Practice
TCC	Texas Chemical Council

1

INTRODUCTION

The Center for Chemical Process Safety (CCPS) of the American Institute of Chemical Engineers publishes a series of *Guidelines* and *Concept* books aimed at defining, evaluating, avoiding and mitigating risks associated with potentially catastrophic events in facilities handling chemicals. Other books were prepared for the purpose of shifting the emphasis on process safety to the design stages, or assisting in understanding fields of technical expertise related to catastrophic failures. This *Concept* book is intended to present a look toward the client–contractor relationship, to assist with delineation of responsibilities and typical splits of work as they influence process safety, and to address indemnification and/or warranties. The purpose of the book is to provide a clearer understanding of the risks that might fall upon each party and how those risks can impact both the business and the relationship.

There are several different types of contractors for which this book is being written: engineering, procurement, and construction contractors (EPC), constructors, consultants, testing firms, contract employees, and miscellaneous subcontractors. Each of these contractors can serve a vital role in the safe design, construction, maintenance, process modifications, risk management, or other aspect of the plant operations. The context from which this book is written is one of addressing regulatory requirements (primarily OSHA Process Safety Management or PSM), process safety design and construction issues, and the ongoing process plant integrity as a joint responsibility that must be accepted by all parties. A key issue is thus recognition of risk (primarily safety, but also business), and who is to manage the risk. Simply stated, risk management is central to a sound contractor–client relationship.

While each party to a contractor–client relationship has specific obligations to the other, both contractually and by regulation, this book holds to the premise that the facility owner has the primary responsibility for providing a safe work place. From that premise, discussion evolves into the issues of risk, risk allocation, and other topics.

The OSHA PSM standard has set a framework for contractor–client responsibilities that will be utilized as a foundation for discussion, as it is the regulatory definition of the relationship. The PSM contractors element, from 29 CFR 1910.119, paragraph h, Contractors, reads:

> **[h] Contractors.** *[1] Application.* This paragraph applies to contractors performing maintenance or repair, turnaround, major renovation, or specialty work on or adjacent to a covered process. It does not apply to contractors providing incidental services that do not influence process safety, such as janitorial work, food and drink services, laundry, delivery or other supply services.
>
> *[2] Employer responsibilities.*
>
> [i] The employer, when selecting a contractor, shall obtain and evaluate information regarding the contract employer's safety performance and programs.
>
> [ii] The employer shall inform contract employers of the known potential fire, explosion, or toxic release hazards related to the contractor's work and the process.
>
> [iii] The employer shall explain to contract employers the applicable provisions of the emergency action plan required by paragraph [n] of this section.
>
> [iv] The employer shall develop and implement safe work practices consistent with paragraph [f][4] of this section, to control the entrance, presence and exit of contract employers and contract employees in covered process areas.
>
> [v] The employer shall periodically evaluate the performance of contract employers in fulfilling their obligations as specified in paragraph [h][3] of this section.
>
> [vi] The employer shall maintain a contract employee injury and illness log related to the contractor's work in process areas.
>
> *[3] Contract employer responsibilities.*
>
> [i] The contract employer shall assure that each contract employee is trained in the work practices necessary to safely perform his/her job.
>
> [ii] The contract employer shall assure that each contract employee is instructed in the known potential fire, explosion, or toxic release hazards related to his/her job and the process, and the applicable provisions of the emergency action plan.
>
> [iii] The contract employer shall document that each contract employee has received and understood the training required by this paragraph. The contract employer shall prepare a record which contains the identity of the contract employee, the date of training, and the means used to verify that the employee understood the training.

[iv] The contract employer shall assure that each contract employee follows the safety rules of the facility including the safe work practices required by paragraph [f][4] of this section.

[v] The contract employer shall advise the employer of any unique hazards presented by the contract employer's work, or of any hazards found by the contract employer's work.

By setting a regulatory standard for the contractor–client (contract employer–employer) relationship, OSHA caused more attention to be given to that relationship. As a result, more clear statements of responsibilities, shared training efforts, and clearer contractual delineation (especially when duties are being delegated) of responsibilities for process safety issues have been developed.

1.1. OBJECTIVE / SCOPE

The objective of this volume is to help readers understand the basics of the contractor–client relationship in order to facilitate the safe design, construction, operations and maintenance of a processing facility with inherently high process safety, equipment integrity and reliability. The book will address process safety issues from the view of assignment of responsibilities and duties. This includes avoidance and mitigation of potentially catastrophic events that could affect people and facilities in the plant or surrounding area. Philosophically, a sound contractor–client relationship is considered one of many safety layers necessary for a safer process facility. Process safety issues affecting operations and maintenance are addressed where relationship choices might impact system reliability. This includes turnarounds or other periods in the life of a facility.

This *Concept* book highlights various types of contractor–client relationships both collectively and by separate chapter presentations. It is clear that choices made early in the relationship can reduce the possibility of arguments and debate over liabilities, while promoting a joint effort to avoid or mitigate large releases and/or the effects of such releases. Example cases, both successes and failures, are included where appropriate to more clearly state a concept.

Specifically, this *Concept* book is only intended to assist the practicing engineer or other professional in understanding and properly defining the basis of his client–contractor relationship, and thus promote a team spirit and harmony of purpose. The ideas presented here are not intended to replace corporate attorneys, contracts personnel, or contract departments. They are likewise not intended to avoid liabilities that might logically accrue due to the failure to adhere to or attempts to circumvent regulations, codes, or technical and trade society standards. The ideas do, however, embrace the concept that contractors provide necessary services to owners (clients) in supporting the ongoing integrity of process plants.

1.2. ORGANIZATION OF THIS BOOK

This *Concept* book has been organized so that the reader may understand key issues, the types of contractor–client relationships into which he or she may enter, and the apportionment of responsibilities to each party.

The chapters are as follows:

Chapter 1. Introduction
Chapter 2. Contractor Safety Programs: General
Chapter 3. OSHA PSM and the EPC Contractor
Chapter 4. Subcontractor Relationships
Chapter 5. Managing Client–Contractor Risk
Appendices

Specific references are listed at the end of each chapter. Industry standards, codes, and regulations are listed at the end of Chapter 1. Additional sources of information are listed in each Chapter under Suggested Reading. It is not the intent of this book to make specific recommendations, but to provide a source of references where the interested reader can obtain more detailed information. A List of Acronyms and a Glossary are provided.

The readings listed at the end of Chapter 1 are good general sources of information on contractor–client relations and chemical process safety. They are recommended for use in combination with the CCPS *Guidelines* and *Concept* books.

1.3. INTRODUCTION TO TERMINOLOGY

Many common terms are defined in the Glossary at the beginning of this book. In following chapters, several definitions will be necessary to clearly deal with the issues. Among the definitions are:

Process Safety Management (PSM): A program or activity involving the application of management principles and analytical techniques to ensure the safety of process facilities. This is typified by the program in 29 CFR 1910.119 Process Safety Management of Highly Hazardous Chemicals.

Risk: The combination of the expected frequency and consequence of a single accident or group of accidents.

Risk Management: The systematic application of management policies, procedures, and practices to the tasks of analyzing, assessing, and controlling risk in order to protect employees, the general public, the environment, and company assets. In following chapters, the term risk management will be used to apply to both process risks and business risks.

In discussing types of contractors, the designations must be clear. Definitions are:

EPC Contractor: A contractor who is able to offer the full range or any portion of engineering design, procurement, and construction activities on a single project utilizing in-company resources.

Constructor: A contractor who performs only construction activities on a project. In this case, engineering design may be performed by the client company or another contractor.

Consultant: Typically a small contractor who may provide services, typically professional advice or opinions, related to any aspect of the plant life cycle.

Subcontractor: A contractor who takes on some or all of the obligations of the primary contractor. Examples of subcontractors may include insulators, electricians, plumbers, or specialized consultants.

Contract Employee: An employee who is employed directly by the client on a contract basis. This type of employee is typically a self-employed independent contractor who is not covered by

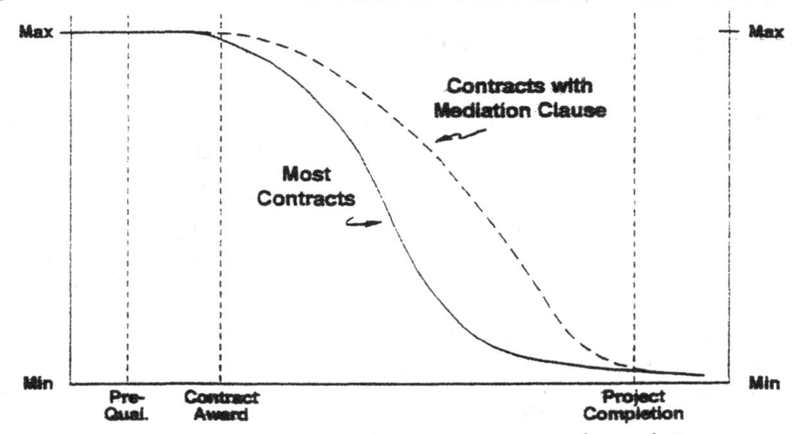

Figure 1.1. Opportunity to improve contractor–client relations.

Subcontractor: A contractor who takes on some or all of the obligations of the primary contractor. Examples of subcontractors may include insulators, electricians, plumbers, or specialized consultants.

Contract Employee: An employee who is employed directly by the client on a contract basis. This type of employee is typically a self-employed independent contractor who is not covered by company benefits programs, but in other aspects such as work hours or duties may appear to be a direct employee.

Another area where terminology may not always be clear is the contractual basis, such as:

Lump Sum: A lump sum project is one for which a single, lump sum payment is made. This type of project is usually of short duration due to the time value of money, but it is not limited in time.

Fixed Price: A fixed price contract is similar in nature to a lump sum contract, but has a payment schedule which provides a periodic cash flow to the contractor.

Time and Materials: Time and materials is similar to fixed rate, reimbursable, in that the contractor's time or worker-hours are paid by the client at an agreed rate; whether at a multiplier of the contract employee's salary, or at a fixed rate. Other contractor expenses are typically reimbursed at cost or at an agreed markup.

Upon review of the terms explained above, the reader may begin to get a sense of the complex issue of assigning or apportioning the risks related to a large petrochemical design and construction project. Each of the various contractors involved will shoulder a portion of the risk based upon his level of participation in the project and his agreement with the client. The client, however, will usually carry the bulk of the risk, since the client also will profit most from a successful project. At the heart of the contractor–client relationship then is a contract that defines the basis and apportions risk among the parties.

On smaller projects or when dealing with contract employees, the client should be aware that he may assume most or all of the liabilities unless he is contractually released from those liabilities and the release is valid and enforceable. An awareness of liability issues and the allocation of responsibilities is discussed in more detail in Chapter 5.

1.4. THE CONTRACT OPPORTUNITY

Having stated that a contract is the basis for apportioning risk among parties, the next logical conclusion is that the contract is the statement of and represents the contractor–client relationship. It is the covenant or binding agreement to which parties will perform. The contractual risk issues should then be addressed as early as possible by all parties to the final agreement, taking advantage of the best opportunity to properly define the relationship.

Figure 1.1 has been prepared as a graphical representation of the opportunity for or effectiveness of open discussion regarding the contractual contractor–client basis. Note that a contract with a mediation clause offers more opportunity for conflict resolution, but the maximum benefits may be obtained only by addressing key issues at the start of the relationship.

Mediation is introduced so that readers may understand that it is one legal alternative (and growing in popularity) for settling contractual differences. The process of mediation occurs in a diplomatic manner with a mediator to facilitate open discussion. Another legal remedy is arbitration by an appointed arbiter who makes a binding decision. Arbitration is more adversarial than mediation, but less costly than a court proceeding. For further information regarding alternatives for resolving contract disputes, the reader is advised to discuss specific issues with their corporate counsel.

Many benefits may be obtained through early resolution of the contractual relationship, primarily through better definition of the work scope and work allocation. Some of the possible benefits include:

- Improved safety awareness at all levels of both contractor and client organizations,
- Improved employee (including contractor employee) knowledge of safety hazards inherent to the process and the hazardous chemicals in the workplace,
- Wider participation in managing the workplace,
- Reduced potential for catastrophic accidents,
- Reduced number and severity of all incidents,
- Improved communication within the plant facilities and with the surrounding community,
- Enhanced implementation of emergency action plans and emergency response plans,
- Quicker and less costly environmental cleanup,
- Improved employee morale from understanding project/company safety protocols and recognition that safety concerns are addressed.

1.5. REFERENCES

1.5.1. Regulations, Codes of Practice, and Industry Standards

29 CFR Part 1910—Safety and Health Standards. Occupational Safety and Health Administration (OSHA).

29 CFR Part 1926—Construction Standards. OSHA.

29 CFR 1910.119. *Process Safety Management of Highly Hazardous Chemicals*. Occupational Safety and Health Administration (OSHA).

40 CFR Part 68. Risk Management Programs for Chemical Accident Release Prevention. Environmental Protection Agency (EPA).

OSHA Instruction CPL 2-2.45A CH-1. September 13, 1994. OSHA Directorate of Compliance Programs.

Responsible Care, Process Safety Code of Management Practices. 1990. Chemical Manufacturers Association (CMA), Washington, D. C.

API (American Petroleum Institute) RP 750. 1990. *Management of Process Hazards.* 1st ed. American Petroleum Institute, Washington, D. C.

API RP 2220, CMA Manager's Guide. 1991. *Improving Owner and Contractor Safety Performance.* 1st ed. American Petroleum Institute, and Chemical Manufacturer's Association, Washington, D. C.

API RP 2221, CMA Manager's Guide. 1995. *Implementing a Contractor Safety and Health Program.* 1st ed. American Petroleum Institute, and Chemical Manufacturer's Association, Washington, D. C.

1.5.2. Suggested Reading

AIChE Annual Engineering and Construction Conference *Panel Presentations*. AIChE, New York.

Journal of Loss Prevention in the Process Industries. Butterworth-Heinemann. London.

Lees, F. P. 1980. *Loss Prevention in the Process Industries.* 2 Volumes. Butterworths, London.

Large Property Damage Losses in the Hydrocarbon-Chemical Industries A Thirty Year Review. Sixteenth Edition, 1995. M&M Protection Consultants, New York.

Loss Prevention Symposium Series. Papers presented at the Annual AIChE Loss Prevention Symposia. American Institute of Chemical Engineers (AIChE), New York.

Process Safety Progress (formerly Plant/Operations Progress). T. A. Ventrone, ed., Quarterly publication of American Institute of Chemical Engineers (AIChE), New York.

Recommended Guidelines for Contractor Safety and Health. 1990. Texas Chemical Council Occupational Safety Committee Report.

John Gray Institute, Managing Workplace Safety and Health: The Case of Contract Labor in the U.S. Petrochemical Industry. 1991. Lamar University System, Beaumont, Texas.

The AMA Project Management Book of Knowledge Series (PMBOK). American Management Association. New York.

The Honourable Lord Cullen, *The Public Inquiry into the Piper Alpha Disaster*. Vols. I & II. 1990. UK Department of Energy, London, England.

2

CONTRACTOR SAFETY PROGRAMS: GENERAL

The evolution of contractor safety programs has historically been based upon industry experiences coupled with the recognition that poor safety is costly in terms of productivity, personnel turnover, capital costs, and insurance. The advent of the Occupational Safety and Health Administration (OSHA) in 1970 served a role by providing specification-based safety practices and procedures for industry, coupled with regulatory oversight. More recently, the evolution has resulted in performance-based standards such as 29 CFR 1910.119, *Process Safety Management of Highly Hazardous Chemicals*, which is addressed in more detail in Chapter 3.

2.1. INTRODUCTION

The contract labor industry has come under scrutiny in recent years (John Gray Institute), with some parties arguing that direct-hire employees are the only appropriate alternative for a safe workplace. Industry needs contractors, however, especially for construction, turnarounds, and unique services (blasting, chemical cleaning, x-rays, vibrational analyses, acoustic testing, etc.). Contractors should be capable of demonstrating a track record and commitment to being on time, safely, at a fair price. Just as continuous improvements are strived for by owners, the contractor must show the same sort of commitment and types of programs. In truth, great strides have been made in contractor safety since the John Gray Institute issued its report, with the efforts of trade organizations and the advent of PSM serving as primary contributors.

> Owners and contractors need to provide safe workplaces and to protect the safety and health of their work forces and the general public. When they work together to improve the safety and health of contractors, both benefit. Benefits that come from a comprehensive and systematic contractor safety program include the following:
>
> a. Safety and well-being of contractor and owner employees are increased.
> b. Improved quality and productivity occur because a comprehensive contractor safety program requires that workers be properly trained for their job tasks and familiar with their job requirements.
> c. Fewer incidents result in less need for regulatory action and more controllable project costs.
> d. The potential for damage to the owner's facility and the owner's equipment is minimized. (API RP 2220)

The above quotation from the joint efforts of the American Petroleum Institute (API) and the Chemical Manufacturers Association (CMA) indicate the general understanding and acceptance of the need for good contractor safety programs. Other industry groups, such as the Associated General Contractors of America (1990) and The Business Roundtable (1982) also address these programs.

In order to properly address these issues, the client should clearly state what is required for safety (including process safety) during the bid process, thereby allowing the contractor to include the necessary services (e.g., site specific training) in his pricing. Also, reviews to confirm the contractor's safe work history should be completed during the bid process to facilitate contractual resolution. This early resolution will allow parties to avoid common differences between corporate legal staff concerns and those of the contractors and plant safety personnel. As an example, contractors and plant safety personnel are charged under OSHA PSM to determinate facts in the event an incident occurs and to disseminate the facts. This may be contrary to the wishes of the client's corporate legal counsel. If the contract addresses this issue at the start of the relationship, there is less confusion.

2.2. SAFE WORK PRACTICES

One key aspect of a sound safety program involves the development of safe work practices and procedures, and training on those procedures. Further, the contractor's (or client's) adherence to those procedures must be strongly enforced to assure the appropriate effect. Most labor union contracts clearly address this importance by providing recourse for dealing with employees who do not comply with safety requirements. One of the most effective motivating factors for compliance with safety practices, however, is peer pressure from personnel who accept and recognize the value of safe practices for themselves and their co-workers.

As a guide for engineers who are unfamiliar with the regulatory bases for safe work practices, Table 2.1 provides a partial listing of required safe work programs and personnel safety regulations, obtained from 29 CFR 1910—Occupational Safety and Health Standards. Contractors need to be aware that they may be required to meet the requirements of both 29 CFR 1910 and 29 CFR 1926—Construction Standards.

TABLE 2.1
Example Safe Work Issues and Personnel Safety: OSHA Safety and Health Standards

	29 CFR 1910
Personal Protective Equipment	
General Requirements	.132
Electrical Protective Devices	.137
Eye and Face Protection	.133
Foot Protection	.136
Hearing Protection	.95
Head Protection	.135
Noise Exposure	.95
Respiratory Protection	.134
Welding	.252–.257
Lockout / Tagout	.147
Hot Work Permits	.119
Confined Space Entry	.146
Fire Watch Procedures	.252
Employee Emergency Plans and Fire Prevention Plans	.38
Hazardous Waste Operations and Emergency Response	.120
Safe Use of Hoses	.158
Employee Alarm Systems	.165
Specific Chemicals	
Acrylonitrile	.1045
Anhydrous Ammonia	.111
Benzene	.128
Sulphur Dioxide	.1000
Asbestos	.1001

Work permit procedures have received a considerable amount of attention over the years due to accident histories, resulting in hot work permits being a specified element within OSHA PSM. Sometimes, these permits are used for impromptu warnings and other safety information transfer to the employee. The work permit procedures, however, should be used to ensure that the job is done properly and is documented. They are not substitutes for training.

Piper Alpha

In 1987, a review of the Piper Alpha production platform in the North Sea identified shortcomings in the work permit program, shift changes, and safety training programs (Cullen, 1990). Safety policies and procedures were in place, but training was believed to be inadequate to assure proper implementation. Additionally, shortcomings in the maintenance of fire protection equipment were identified. In 1988, these identified shortcomings are believed to have contributed to the Piper Alpha disaster that resulted in the loss of 167 lives.

The incident which occurred involving the Piper Alpha was one of an ignited gas condensate vapor cloud exacerbated by domino (or knock-on) effects. At the heart of the cause, it is believed, was a lack of proper training on safe work practices. Following shutdown of a pump for maintenance, a safety relief valve was removed (also for maintenance) and the flange was blanked, but not leak-tight. When the parallel pump tripped, the pump being maintained was put into service, resulting in the flange leak believed to have caused the problem. In *The Public Inquiry into the Piper Alpha Disaster*, findings included:

- Failure to transfer information properly at shift change,
- Failure of the work permit program,
- Diesel firewater pumps were on manual start, meaning that they could not be started automatically during a loss of power,
- No formal hazard assessment was required by regulators,
- Safe havens did not exist,
- Personnel did not appear to understand that the personnel quarters (accommodations) should be evacuated.

The Piper incident resulted in several recommendations intended to improve North Sea platform safety. Among these were:

- Better work permit programs,
- Better information transfer at shift change,
- Better fire protection systems,
- Better means of escape (egress),
- Improved standby fleets,
- Periodic audits,
- More regulatory oversight (as opposed to self-regulation), and
- Suggestions for performance-based programs instead of specification-based (which can be self-limiting).

As seen, the Piper Alpha incident investigation points out clearly that the statements at the beginning of this section regarding safe work practices, procedures, training, and enforcement are valid. Hopefully, there will be no need for another such incident to make us aware of the potential consequences.

Other safe work program issues which contractors should address include substance abuse programs, which must be accompanied by testing, and fitness for duty (CMA, 1990). Language issues also need to be addressed, with the contractor and client understanding the attendant risks of a multilanguage workplace, and possibly providing schooling where appropriate.

Construction Issues

Sometimes the client takes on the role of construction manager at the site, overseeing the contract construction personnel. This provides a management oversight that imparts the client's philosophy and procedures, but once again training of personnel must be specific to the client's safe work programs. Table 2.2 provides a listing for construction-related safety issues.

When the owner (client) takes over the position of construction management, he must be careful that all appropriate obligations of a direct employer of the construction workers have been fulfilled. This should be part of the development of a contractual relationship among the parties. For more information, see Chapter 5.

TABLE 2.2
Example Construction Safety Issues—
OSHA Construction Standards

	29 CFR 1926
Confined Space	.21
Excavation and Trenching	.650
Respiratory Protection	.103
Hearing Conservation	.101
Environmental Loss Control	.50–.57
Electrical Work Practices	.400–.499
Grounding	.404
Electrical Power Distribution	.950–960
Battery Charging	.441
Work Platforms	.550
Stairways and Ladders	.1050–.1060
Scaffold Tagging	.451
Hoisting and Rigging	.251, .550, .952
Mobile Equipment	.600–.601
Specific Chemicals	
Nitrogen	.32, .1651
Asbestos	.58

2.3. CONTRACTOR QUALIFICATION

In *Improving Construction Safety Performance: A Construction Industry Cost Effectiveness Project Report*, The Business Roundtable clearly showed that the number and severity of accidents can be reduced by choosing a safer contractor. The study also stated that while OSHA 200 forms (used for numerical tracking of reportable incidents) serve a purpose for transmitting data in a meaningful fashion, other factors should be considered. These can include experience modification rates for worker's compensation insurance, but also should include a review of the contractors' safety attitudes and practices.

When promulgating its process safety management (PSM) standard in 1992, OSHA required companies having covered processes to implement programs to obtain and evaluate data regarding contractors' safety and health programs and performance. Interestingly, this has identified many deficiencies among contractor's programs, but also has bolstered the relationships between contractors and client companies through joint development of program improvements, better communication and training. By working together, contractors and client companies have often become like partners, with common workplace safety goals.

API Recommended Practice (RP) 2221, *Implementing a Contractor Safety and Health Program*, was developed specifically for application with on-site contractors and subcontractors. The RP discusses program elements for implementing contractor safety that include:

- Pre-qualification,
- Selection of Contractor,
- Pre-job Activities,
- Work-in-Progress,
- Evaluation.

API RP 2221 also makes the point that some evaluation of the level of risk (or relative risk) to on-site employees should be made, and suggests qualitative means by which this may be accomplished. The risk issues are expanded to include process safety issues, from which the client may ascertain the level of his involvement with contractor safety programs or how comprehensive a program to require of his contractors.

Most clients now have formal programs in place for reviewing contractor safety and health programs. These include evaluation and qualification of existing contractors as well as pre-qualification for new contractors. The Appendix of this book contains example checklists for such analyses. Other examples are contained in API RP 2221 Appendices, including:

Appendix A: Sample [Contractor] Pre-qualification Form (PQF)
Appendix B: Sample Cover Letter (To accompany PQF)
Appendix C: Sample PQF Evaluation Form
Appendix D: Sample Regulatory Requirements for Selected Work Activities
Appendix E: Sample Contractor Safety and Health Training Matrix
Appendix F: Sample Site Visit Checklist
Appendix G: Sample Contractor Monthly Safety and Health Performance Report
Appendix H: Sample Job Site Inspection Checklist

Appendix I: Safety and Health Evaluation Form
Appendix J: Sample Contractor Safety and Health Policy
Appendix K: Sample Owner's Contractor Safety and Health Written Program
Appendix L : Sample Audit Protocol for Owner's Contractor Safety and Health Program

Commitments by management to safety program quality are also seen as necessary, just as with any other comprehensive corporate program. "Management accountability for safety performance is a very important factor in determining a company's safety record. Companies that hold their project management accountable for accidents along with productivity, schedules, quality, etc., are the ones which have the best safety records." (Business Roundtable, p. 19)

Contractors should be required to comply with all client safety program requirements during the bid/proposal period or else the contractor's bid should be deemed invalid or nonresponsive. Contractors should also show successful compliance with regulatory mandates in the past. On existing contracts, the contractor must show good safety performance, or no contract extensions should be given. This includes documentation that levels or degrees of training for contract personnel are sufficient for the job skills that the project demands.

When contractors are in the plant on a consistent basis, there is continuity of site knowledge, training, and personal understandings and relationships. This requires a continuing commitment from the client. Partnerships have addressed this issue somewhat, but the client commitment is sometimes dependent upon quarterly financial reports. Monetary incentives sometimes have been used for safety with positive results reported. (Texas Chemical Council, 1990)

Should a contractor identify safety problems at a client's facility, he should tell the client (owner) of the problems and request that the client respond (this requires a two-way commitment). Anything less than an appropriate response from the client, however, should result in the contractor considering withdrawal of his employees from the site. As always, the client (owner) has the primary responsibility for providing a safe workplace, but it must be shared by the contractor.

2.4. DOCUMENTATION

Documentation is as important to long-term facility management as it is to the day-to- day safe operation of a chemical facility. Additionally, failure to comply with regulatory mandates to maintain accurate and complete records can become a serious legal liability. Since documentation is the means to implement and verify process safety programs and program compliance, the contractor and the client must work together to define the documentation procedures for a project and closely adhere to those procedures.

Many industry guidelines (e.g., API RP 750 and the CMA Process Safety Code) and regulations (e.g., 29 CFR 1910.119, 40 CFR Part 68, California Risk Management & Prevention Program legislation, New Jersey Toxic Catastrophe Prevention Act) dictate document requirements and retention periods. The *Guidelines for Engineering Design for Process Safety* (CCPS, 1992) emphasize that process safety issues touch on all aspects of plant design, operations and maintenance, and that each of these issues requires documentation. Process safety depends on how a unit is designed, constructed, operated and maintained. A permanent record of the design basis and operational requirements through a document (or records) management system is therefore essential, and is a shared responsibility of the contractor and the client during the course of their relationship. The primary elements of a document management program are procedures, retention and control. (CCPS, 1992)

In *Guidelines for Process Safety Documentation*, CCPS has addressed most aspects of documentation relating to process safety, including records management, contractor issues, and documentation throughout the plant life cycle. Essentially, the book is dedicated to the philosophy that proper documentation should be developed and maintained which defines and represents the status of the facility at each stage in its history. Contractors must be aware that they can be involved in any or all of those stages, and that they have a responsibility to support the documentation efforts.

A sound document management system will require that a document purge procedure for out-of-date information be in place. Examples where purging of documents might be appropriate include:

- OSHA 200 logs are required to be maintained for five years; purging after six years might be suggested.

- Work permits are generated daily, or on a shift basis; purging the prior week's permits daily might be suggested (note that PSM might require these for an incident investigation, but they would be available for more than the necessary 48 hours).
- The two most recent PSM Compliance Audits must be retained; others might be purged.
- Incident investigation reports must be maintained for five years; purging after six years might be suggested.
- Management of change documents must be retained so long as they are necessary for a specific purpose; purging of these documents, which can become numerous, should be reviewed based on the specifics of a company's Management of Change (MOC) procedures.

Document retention and purging is a very important topic due to the interplay of regulatory requirements (encouraging or requiring retention) and legal liabilities (sometimes encouraging purging). There is also an issue of storage space and access if documents are retained in paper format. A clear definition of each document's purpose, the information it contains, and whether there is a need for permanent retention should be addressed before assigning a retention/purge schedule. Retention/purge schedules are also appropriate when records are retained electronically, since electronic (computerized) storage merely improves storage capacity and access. Ideally, these issues will be institutionalized through a corporate records retention program.

A Records Retention Example

In 1994, a company in Louisiana decided to perform an initial PSM audit to define its current compliance with PSM. A consultant was brought in to review the programs in place and advise company management of the status. A key portion of the audit was to be a records review, because the company understood that its program was not yet complete and needed advice on how to proceed. Interestingly, the company had retained an unusually high percentage of all of the documents pertaining to its forty-plus year history. These documents even included the original design calculations for portions of the plant. Unfortunately, the design contractor did not clearly identify which of his three design cases was the final basis.

Similarly, many equipment drawings were filed, often without drawing dates. The result was that a considerable field effort was required to determine which, if any, of the drawings could be utilized by the company in setting up its equipment history files under the mechanical integrity element of PSM.

2.5. REFERENCES

2.5.1. Specific References

29 CFR 1910.119. *Process Safety Management of Highly Hazardous Chemicals*. Occupational Safety and Health Administration (OSHA).

John Gray Institute, *Managing Workplace Safety and Health: The Case of Contract Labor in the U.S. Petrochemical Industry*. 1991. Lamar University System, Beaumont, Texas.

API RP 2220, CMA Manager's Guide. 1991. *Improving Owner and Contractor safety Performance.* 1st ed. American Petroleum Institute, and Chemical Manufacturer's Association, Washington, D. C.

AGC Guide for a Basic Safety Program. 1990. Associated General Contractors of America, Washington, D.C.

Improving Construction Safety Performance: A Construction Industry Cost Effectiveness Project Report. 1982. The Business Roundtable, New York.

29 CFR Part 1910—Safety and Health Standards. Occupational Safety and Health Administration (OSHA).

29 CFR Part 1926—Construction Standards. OSHA.

The Honourable Lord Cullen, *The Public Inquiry into the Piper Alpha Disaster*. Vols. I & II. 1990. UK Department of Energy, London, England.

Responsible Care, Process Safety Code of Management Practices. 1990. Chemical Manufacturers Association (CMA), Washington, D. C.

API RP 2221, CMA Manager's Guide. 1995. *Implementing a Contractor Safety and Health Program.* 1st ed. American Petroleum Institute, and Chemical Manufacturers Association, Washington, D. C.

Recommended Guidelines for Contractor Safety and Health. 1990. Texas Chemical Council Occupational Safety Committee Report.

Guidelines for Engineering Design for Process Safety. 1993. AIChE–CCPS. New York.

Guidelines for Process Safety Documentation. 1995. AIChE–CCPS. New York.

2.5.2. Suggested Readings

AIChE 24th Annual Engineering and Construction Conference Panel Presentations, *Managing Change in the 21st Century*. September, 1992. San Francisco, California.

AIChE 25th Annual Engineering and Construction Conference Panel Presentations, *Today's Challenges—Today's Successes.* September, 1993. Arlington, Virginia.

Richard D. Hislop, *A Construction Safety Program*. September, 1991. Professional Safety.

3

OSHA PSM AND THE EPC CONTRACTOR

This chapter addresses a myriad of issues relating to compliance with 29 CFR 1910.119 *Process Safety Management of Highly Hazardous Chemicals* (OSHA PSM). The template chosen as a basis for this discussion is presented as Table 3.1 in Section 3.3. Each element of PSM is discussed in Section 3.2 in order to provide a clearer understanding of the choices represented by the table.

In reviewing Table 3.1, it is important to remember that the table is intended to address the activities and shared responsibilities between an Engineering–Procurement–Construction (EPC) Contractor and his client. The allocation of responsibilities might be expected to be considerably different when defining a relationship between other types of contractors and the same client company. Interestingly, one chemical company uses Table 3.1 in modified forms (1) for capital contracts (basically as adopted), (2) for in-house designs, and (3) as a basis for the split of work between departments for PSM development.

For purposes of clarification, it is useful to understand the major activities historically included in engineering, procurement and construction. These include (Kerridge, p.155):

Engineering: Process design, conceptual/analytical design, production design, specifications, requisitions and drawings.

Procurement: Inquiries, bid evaluations, purchase orders, expediting, inspection and delivery to job site.

Construction: Temporary facilities, material receipt, and erection of civil, structures, equipment, piping, instruments, electrical, paint, and insulation.

Within this framework, an example definition of project PSM activities and a possible split of responsibilities for those activities is discussed below.

3.1. PROJECT PLANNING UNDER OSHA PSM

Project planning is complex on every substantial project, but can be exacerbated without a clear definition of the activities to be performed under each regulatory requirement and a specific assignment of responsibilities for those activities. In Section 3.2, each element of OSHA PSM is addressed, and commentary regarding the choices of assignment of responsibility is provided.

Often, it is assumed that the PSM-related engineering activities are clear for an EPC contractor because most of the necessary activities are historically present in a sound design. Without a "road map," however, designers cannot be expected to provide a product that fits into the client's PSM program. Remember, the scope of work required for a PSM-covered process has evolved and expanded, and the necessary activities need to be delineated and assigned to the appropriate party.

Procurement activities, along with construction activities, can be the linchpins between the engineering design intent and the final product as constructed. For procurement, PSM-related activities may include:

- Assuring suitability of equipment suppliers and subassembly equipment suppliers,
- Preparing a pre-approved contractors (or subcontractors) list,
- Assuring that purchase requisitions address common safety related issues prior to being released for fulfillment,
- Bringing appropriate process safety personnel to vendor review meetings before releasing final purchase orders,
- Performing appropriate shop inspections to confirm that appropriate codes, standards and accepted engineering practices are followed,
- Expediting suppliers to avoid out of sequence deliveries, and thus avoiding out of sequence construction activities,

- Assuring that vendors supply the necessary standard operating and maintenance procedures for equipment,
- Assimilating and distributing vendor drawings, prints and other pertinent information so that equipment data files may be comprehensive.

Many PSM-related activities are necessary for construction to perform, including the following:

- Assisting in establishing and applying applicable codes, standards and good engineering practices in the construction efforts,
- Assuring that construction is in accordance with the design of new equipment (through field inspections, P&ID checks, punch lists, pre-startup safety review [PSSR] support, etc.),
- Documenting assurance of new equipment materials of construction (through confirmation of materials received, field checks versus P&IDs, positive material identification or other means, etc.),
- Assisting in documenting that the new equipment adheres to the design specifications and manufacturer's installation instructions (through confirmation that design specifications and manufacturer's instructions are properly implemented during a PSSR),
- Documenting any tests that were performed to assure that the equipment was suitable for its design intent and properly installed (through hydrotest sheets, electrical checks, instrument checkout lists, calibration sheets, etc.),
- Following all applicable client safe work practices when performing construction activities, especially when near operating equipment for an existing covered process,
- Assisting in assurance that subcontractors are also following the appropriate safe work practices (through use of safety analysis forms, job safety analysis forms, etc.),
- Assisting engineering and the client as necessary in performing a PSSR before the initial startup of newly constructed equipment or facilities (or modified facilities that involve construction),
- Enhancing the transfer of information from construction to the client's maintenance department to ensure that the documented equipment history begins with fabrication and includes inspection and construction information.

3.2. PSM ELEMENTS

A general issue for each project that may be covered under OSHA PSM is that of defining the PSM project plan basis. Section 3.3 suggests that a project PSM basis be prepared for each project, and Table 3.1 is intended to offer a starting point for discussion. As pointed out in the description of the compliance issues, it is important not just to have a plan, but also to clearly define the scope of activities under the plan, and the boundaries of the covered process.

Employee Participation

The employee participation element of PSM is typically regarded as a client activity and responsibility. Consequently, Table 3.1 suggests that the client be responsible for this element of PSM. The contractor may have certain responsibilities under the element, however, such as leadership of a process hazard analysis, or other aspects of the PSM program. For ongoing, on-site contractor relationships (e.g., maintenance contracts), the contract employees should be included in the employee participation program as if they were direct employees.

Process Safety Information

Developing content for the process safety information (PSI) element of PSM is primarily regarded as the responsibility of the EPC contractor on most projects. Specific material safety data sheets may be the sole responsibility of the client due to the attendant liability that might accompany improper or inaccurate data. Other PSI, however, is typically the design basis and design product from the EPC contractor. Most EPC firms have a standard data book which they offer on all projects, and this should be reviewed as a standard contractual issue since the client must ultimately assume the responsibility for keeping PSI current. Such a data book, following OSHA's requirements, might include the following:

Information Pertaining to the Technology of the PSM Process, including:

- Block flow diagram or simplified process flow diagram (PFD).
- Process chemistry.

- Maximum intended inventory. In batch operations and for many vessels in continuous processes, this is the same as the maximum design capacity for the specific tank or vessel.
- Safe upper and lower limits for items such as temperatures, pressures, flows or compositions. Often, the limits may be the same as design limitations on the equipment.
- An evaluation of the consequences of deviations from safe upper and lower limits, including those affecting the safety and health of employees.

Information Pertaining to the Equipment in the Process:

- Materials of construction.
- Piping and instrumentation diagrams (P&IDs).
- Electrical classification drawings.
- Relief system design and design basis.
- Ventilation system design.
- Design codes and standards employed.
- Material and energy balances.
- Safety systems.

Material and energy balances are most commonly utilized to represent continuous process operations that are more readily modeled mathematically and where an energy balance is an integral part of the process design. For certain batch operations where the client deems it inappropriate, or where an energy balance has no impact upon the process, the PSM requirement for a material and energy balance may be considered to be sufficiently met through the preparation of a material balance only, but documentation of such an agreement is critical to the contractor–client relationship.

Process Hazard Analysis

When scheduling a process hazard analysis (PHA) on a capital project, a cursory view is often taken that a PHA will be scheduled once the available information is sufficient. From this point, recommended mitigative actions will be incorporated into the final design. If such an approach is to provide confidence of successful mitigation, there must be controls to implement and document resolution of appropriate mitigative actions.

An appropriate project PHA is one part of a comprehensive project safety program that will include plot plan safety reviews, preliminary hazard analyses, pre-PHAs, implementation of management of change, confirmation of the "as-built" status of piping and equipment, procurement and construction quality assurance, pre-startup safety reviews, and possibly a follow-up PHA prior to plant startup. The methodology chosen for hazard identification may vary based on the timing of any study, the purpose or intent of the study, or even availability of personnel. For more guidance on hazard evaluation, it is suggested that the engineer review *Guidelines for Hazard Evaluation Procedures, 2nd Edition with Worked Examples*.

Operating Procedures

Operating procedures are typically unique in form and content for each operating company. Unfortunately, many operating companies lack the resources to develop procedures for major EPC process projects. Most large engineering firms have prepared in-house policies for developing project PSM programs and process safety information manuals, and these may be used to assist in developing operating manuals.

Despite the choice of form and content, the operating procedures must be written to address several issues as required by PSM, including:

- Steps for each operating phase, including:
  —Initial startup;
  —Normal operations;
  —Temporary operations;
  —Emergency shutdown (including the conditions under which required, and the assignment of shutdown responsibility);
  —Emergency operations;
  —Normal shutdown; and,
  —Startup following a turnaround, or an emergency shutdown.

- Operating limits, including:
  —Consequences of deviation; and
  —Steps required to correct or avoid deviation.

- Safety and health considerations, including:
  —Properties of, and hazards presented by, the chemicals used in the process;
  —Precautions necessary to prevent exposure, including engineering controls, administrative controls, and personal protective equipment;
  —Control measures to be taken if physical contact or airborne exposure occurs;
  —Quality control for raw materials and control of hazardous chemical inventory levels; and,
  —Any special or unique hazards.
- Safety systems and their functions.

The operating procedures must also be kept current, reflecting current operating practices and information on process chemicals, technology, and equipment. In order to fulfill this responsibility, the owner might have the contractor prepare and document the "as-built" status at the completion of the project. Under all circumstances, it is imperative that all parties understand the status of documents when the client takes custody.

The operating procedures element also requires the development and implementation of safe work practices to provide for the control of hazards during operations such as lockout/tagout; confined space entry; opening process equipment or piping; and control over entrance into a facility by maintenance, contractor, laboratory, or other support personnel. The safe work practices will apply equally to client employees, contractor employees, or others (such as subcontractors and vendors) who might provide services where they will come into contact with the process operations.

Training

Training is often performed by client companies such that a corporate philosophy of form and content is apparent. For those companies (typically larger corporations), the EPC contractor will only supply information to the trainers. For smaller client companies or for new technologies, however, the initial training of operators and other key personnel may fall upon the EPC contractor.

In addition to operating procedures, training considerations include training on safe work practices, emergency action plans, process overviews and other topics besides basic job skills. Recent trends include annual job skill certification for crafts being performed and certified by local area contractor safety councils. Another trend finds companies providing all safe work practices, emergency action plans and process overview training directly to contractor personnel in order to assure and document that the proper training has been performed.

When the EPC contractor is providing or supporting training efforts, it will typically represent the initial training required under PSM. Training should be provided to all employees who might be involved in operating the process, and should be based upon the operating procedures discussed above. Emphasis should be placed on the specific safety and health hazards, emergency operations including shutdown, and safe work practices applicable to the employee's job tasks. Documentation of the training should also be kept to meet the requirements of PSM. The documentation should include documentation of the qualifications of the instructors.

Contractor Programs

The contractor safety requirements stated under OSHA PSM are categorized as either the contractor's responsibility or the employer's (client's) responsibility. Interestingly, in practice both parties must share in the efforts to meet the requirements. If a contractor does not meet its obligations, then the client company does not have a safe workplace. Conversely, if the hazards of the workplace are not clearly identified and explained to the contractor's workers, then the client again has an unsafe workplace. No better example of the need for cooperation exists than in the case of emergency action plans and emergency response and responder training. Should the contractor be involved in responding to an emergency, the contractor must become as one with the client's response team.

In its Compliance Directive, Instruction CPL 2-2.45A CH-1, OSHA clarified many issues relating to contractors, contract employees, and other topics of concern. This clarification represented a legal settlement with labor unions. Of utmost interest to EPC firms, as well as clients and other contractors, is the resulting philosophy that focused more attention on contractor employees. Essentially, contract employees must be treated the same as direct hire employees for the task at hand, thereby expanding training requirements and contract employee participation in other elements. This is also very evident from a cursory review of Appendix A "PSM Audit Guidelines" and the enclosed audit checklists.

In 1994, a chemical release occurred at a test facility of the Lyndon B. Johnson Space Center near Houston, Texas. Because NASA had historically been considered exempt from OSHA compliance, NASA had not performed the required employer's duties under OSHA's PSM. This was probably exacerbated due to the perceived need of NASA for security. Unfortunately, several of NASA's contractors on-site were not aware of the potential hazards of the chemicals present and did not have emergency action plans in place for their employees. If NASA had conveyed the nature of the chemical hazards and had prepared an emergency response plan that included (and trained) contractors, the near miss would not have become a public relations fiasco.

Pre-Startup Safety Review

Pre-startup safety reviews are an important issue for the contractor and client to resolve. Pre-startup audits follow on the heels of mechanical completion and pre-commissioning activities, and the split of responsibilities will depend on other contractual obligations. For example, if pre-commissioning and startup is to be performed by the client company, an EPC contractor assumes unwarranted liabilities by merely adding lubricants to rotating equipment. Conversely, if the EPC contractor is to start the plant, a client employee who modifies a control station or distributed control system function might nullify process guarantees. Each of these issues must be addressed contractually if the relationship is to remain sound. And unsound relationships exacerbate communication and may lead to failures in management systems and process safeguards.

Generally, client companies and contractors utilize checklists to ascertain and document equipment status prior to startup. If such a checklist is to play a role in the custody transfer of completed facilities (from the contractor to the client), it should be included in the contractual basis of the project.

Mechanical Integrity

The mechanical integrity program is probably the PSM element for which operating companies require the most contractor support and expertise. The specialized nature of many of the testing and inspection procedures requires up-to-date certification on state-of-the-art techniques, which most operating companies use infrequently. Some EPC contractors have the expertise necessary to employ appropriate techniques for baseline data development, but others have to rely on equipment vendors and subcontractors. A delineation of responsibilities for the mechanical integrity program baseline data gathering and development thus becomes more important. For example, who is to certify field welds on pressure vessels or bench test results for safety relief valves? For safety-critical electronic control systems, does the factory acceptance test suffice, or is there a witnessed test in the control room prior to plant startup? Close cooperation between the operating company's quality assurance personnel and those of the EPC contractor and its subcontractors is crucial.

For older existing plants, where an EPC expansion project must tie in with existing equipment, correction of equipment deficiencies under paragraph (j)(5) can become a major problem. As stated, the correction of deficiencies must meet the requirements defined by paragraph (d), process safety information. Those requirements are:

> For existing equipment designed and constructed in accordance with codes, standards, or practices that are no longer in general use, the employer shall determine and document that the equipment is designed, maintained, inspected, tested, and operated in a safe manner.

Issues that arise in this instance include understanding of the original (out of date) codes and standards, whether equipment revisions/upgrades must meet the latest edition of codes and standards, and contractor qualifications to make any appropriate modifications. In order to be certain, clients should ask contractors to define and document the necessary qualifications, as well as work closely with the contractor to ensure that administrative controls which have resulted in prior safe operations are fully understood and incorporated.

A final issue to note under the mechanical integrity element is the need for written procedures for maintaining equipment integrity. As stated earlier in this chapter, the procurement activities might include "assuring that vendors supply the necessary standard operating and maintenance procedures for equipment." Whenever the equipment is particularly unique in form or function, or whenever it is a type of equipment that is new to the plant site, then development of the written procedures and the proper training of personnel take on added significance.

Hot Work Permit

Hot work permits are considered to be a simple (and standard) safeguard for employee safety, but failures in permitting systems are commonplace. Many times, permitting system breakdowns occur because employees are attempting to follow one party's (contractor's) procedure when the other party's (client's) procedure is required. Other simple mistakes involve fire watchers, signatures, clear instructions, and shift changes. Training on permitting systems is necessary for all involved. As stated in Chapter 2, the work permits should not be used as a means to transfer information that should be addressed by training programs. They might, however, be used to confirm that the training is in place.

Management of Change

Management of change is a difficult program to define or establish when dealing with the EPC contractor–client interface. It is certainly easy to say that all changes beyond a certain point in the design are to be covered by management of change (MOC) procedures, but in practice this is a very complex issue. In fact, for each portion of the EPC effort, different rules may apply.

During the engineering design process, most changes occur in normal fashion within the engineers' revision control procedures. These typically have considerable flexibility and are not invoked on a project until absolutely necessary. The reason is that changes of pipe or equipment sizes, specifications, or other considerations that are influenced by fluid hydraulics or process fluctuations are by nature iterative during design. Excessively rigid controls at this stage do not allow for the normal design interplay. Historically, contractors have established controls that became more rigid as the project progressed, and these have usually served the client well. Management of change for the design process is then most often implemented at the completion of the process hazard analysis, where tracking of recommended mitigation actions is a regulatory necessity. At that time, the final stages of the revision control system can be converted to the client's management of change program. The changeover from the revision control procedures to the client's MOC procedures must be a well-defined process, and key personnel should be trained to ensure proper implementation. If the client's MOC procedure is not properly implemented, changes might be implemented without proper safety reviews and documentation. This compromises both the MOC procedure and the entire PSM program.

In the procurement efforts, management of change should be ingrained in purchasers, expeditors, and inspectors. The need for pre-qualified vendors, strict vendor adherence to the established engineering codes, standards and practices, company (client) guidelines, and other potential concerns such as material identification or verification must be understood and practiced by all. The procurement cycle thus becomes a critical path item for the PSM program.

Construction is also a critical stage where management of change procedures must be closely followed. Should field modifications not be transmitted back to engineering for a hazard assessment, nasty surprises could be encountered at startup. Something as simple as a piping material specification change on the wrong side of a pressure let down valve could result in brittle fracture of inappropriate materials, leading to a major loss of equipment and possibly lives.

Incident Investigation

OSHA requires that contractors be included on incident investigation teams when an incident occurs that involves contractor employees. An EPC contractor, when on-site, should assign a site PSM coordinator who is identified as the incident investigation team member. This will allow the client to train that person as part of the team, and expedite investigations should they involve contractor employees. One chemical company has a stated policy that any contractor who does not agree (prior to the start of his contract) to be part of an incident investigation team when asked, will not be allowed into the plant. This reflects the importance assigned to incident investigation by many companies.

Emergency Planning and Response

Emergency action plans (including training of contractor employees on the plans), alarms, and notification procedures are the heart of the emergency planning and response element. Typically, client personnel are trained as responders (under 29 CFR 1910.120), and contractor personnel evacuate the premises during any major incident. At a large construction site, contractor personnel could be used as responders, but training and training certification often are cost prohibitive or ineffectual.

Compliance Audits

The PSM compliance audit, which is required at least every three years, is typically the sole responsibility of the client company. Contractor personnel are interviewed as part of any comprehensive site audit, however. Training of contractor personnel on PSM issues, job skills, and safe work practices are issues of concern, along with documentation of any training received.

Trade Secrets

Trade secret information and confidentiality agreements are handled by mutual contractual agreements between companies. Since extension of these agreements to client employees, EPC contractor employees and subcontractors is a separate and complex issue, it will not be addressed in detail within this book.

3.3. CHECKLIST BASIS FOR PSM PROJECT PLAN

Table 3.1 on the following pages represents one company's approach to a work allocation on process safety management (PSM) issues for capital projects. The table presents a common division of work, but within the company's PSM program it is understood that each project will have variations in work assignments. Thus, the following table might represent a starting point for project discussion and PSM work assignments. Note that the example issues in the table represent project commitments that go beyond OSHA's requirements.

TABLE 3.1
EPC PSM Plan Checklist (Example with common allocation of work identified)

Description of PSM Compliance Issues for Capital Projects under 29 CFR 1910.119	PSM Requirement	Project Requirement	EPC Supply	Client Supply
GENERAL ISSUES				
1. Written PSM overall project plan		Y	J	J
2. Assign PSM coordinators for project		Y	Y	Y
3. Split of PSM responsibilities		Y	J	J
4. List of covered chemicals		Y	Y	
5. Description of boundaries of covered process		Y	Y	
6. Provide PSM training to appropriate project personnel		Y	Y	Y
7. Pre-qualification of equipment vendors		Y	J	J
8. Pre-qualification of contractors and subcontractors	Y	Y	J	J
9. Schedule periodic project PSM reviews		Y	J	J
10. Other:				
EMPLOYEE PARTICIPATION—PARAGRAPH C				
1. Written plan for implementation	Y	O	O	Y
2. Documentation of employee involvement in PSM program development and PHA	Y	N	—	Y
3. Other:				
PROCESS SAFETY INFORMATION—PARAGRAPH D				
1. Material Safety Data Sheets	Y	Y		Y
2. Process Safety Information manuals to comply with 29 CFR 1910.119, paragraph [d]	Y	Y	Y	
3. As-built P&IDs, PFDs, mass and energy balances	Y	Y	Y	
4. Equipment data files, as-built interlock diagrams, electrical classification drawings, relief system design basis, other information	Y	Y	Y See 2.	
5. Materials of construction		Y	Y	
6. Documentation that equipment is in accordance with design specifications and that design complies with generally accepted good engineering practice	Y	Y	Y See 2.	
7. Other:				

Description of PSM Compliance Issues for Capital Projects under 29 CFR 1910.119	PSM Requirement	Project Requirement	EPC Supply	Client Supply
PROCESS HAZARD ANALYSIS—PARAGRAPH E				
1. PHA methodology acceptable to OSHA	Y	Y	J	J
2. Performance and tracking of PHA to ensure that entire covered process has been studied, recommendations have been approved and incorporated in design	Y	Y	Y	A
3. Documentation that appropriate team members attended the PHA	Y	Y	Y	
4. Documentation of PHAs retained for project/ process history	Y	Y	Y	
5. Approvals of team leader and team members		Y	Y	Y
6. Documentation that engineering and administrative controls have been considered as well as human factors, facility siting and health effects	Y	Y	Y	
7. Final report to be suitable for company archives		Y	J	
8. Other:				
OPERATING PROCEDURES—PARAGRAPH F				
1. Written procedures for all plant operating modes	Y	O	O	Y
2. Written standard operating procedures obtained for all vendor supplied rotating equipment and packaged units		Y	Y	
3. Safe work practices for control of hazards during nonroutine work activities	Y	Y	O	Y
4. Other:				
TRAINING—PARAGRAPH G				
1. Training on operating procedures provided to all employees involved in operating the process	Y	Y	O	Y
2. Documentation of training for all employees involved in operating the process	Y	Y		Y
3. Other:				
CONTRACTOR PROGRAMS—PARAGRAPH H				
1. Evaluation of contractor's (and subcontractors) safety programs	Y	Y	Y	
2. Company review of contractor's safety programs and performance	Y			

Description of PSM Compliance Issues for Capital Projects under 29 CFR 1910.119	PSM Requirement	Project Requirement	EPC Supply	Client Supply
3. Generic contractor safety manual listing safe work practices to be followed while performing work on (or near) a covered process		Y	J	J
4. Information related to surrounding processes, if any, the safety hazards which might be encountered from them, and emergency actions which may be required	Y	Y		Y
5. Information related to the hazards of the process	Y	Y	J	J
6. Control of entrance/exit of contractors and subcontractors from covered process	Y	Y	J	J
7. Emergency action plan training provided each contractor employee	Y	Y	J	J
8. Documentation of training of contractor employees on hazards and emergency information	Y	Y	Y	
9. Documentation of actions when contractor employee safety infractions are encountered		Y	Y	
10. Injury and illness logs for contractor employees	Y	Y	Y	
11. Process safety topics included in "tool box" safety training		Y	J	J
12. Procedure to advise client of any unique hazards identified by the contractor	Y	Y	Y	
13. Other:				
PRE-STARTUP SAFETY REVIEW (PSSR)—PARAGRAPH I				
1. PSSR procedures	Y	Y	O	Y
2. PSSR forms and checklists in place		Y	O	Y
3. Documentation of PSSR and authorization to start up		Y	O	Y
4. Other:				
MECHANICAL INTEGRITY—PARAGRAPH J				
1. Written procedures to maintain the ongoing integrity of process equipment	Y	Y	O	Y
2. Training of maintenance personnel appropriate to meet the intent of 29 CFR 1910.119, paragraph [j] (site safety, emergency plans, safe work practices)	Y	Y		Y
3. Determination of appropriate frequencies of inspections and tests for covered process equipment	Y	Y	Y	

Description of PSM Compliance Issues for Capital Projects under 29 CFR 1910.119	PSM Requirement	Project Requirement	EPC Supply	Client Supply
4. Documentation of initial inspections and tests and correction of identified equipment deficiencies prior to startup	Y	Y	Y	
5. Quality assurance procedures for new equipment and for maintenance materials, and spare parts	Y	Y	Y	
6. Documentation and certification of training of contractor employees on appropriate craft skills	Y	Y	Y	
7. Obtain manufacturer's installation and repair manuals	Y	Y	Y	
8. Other:				
HOT WORK PERMIT—PARAGRAPH K				
1. Hot work permit procedures and permit forms shall be in place	Y	Y		Y
2. Other:				
MANAGEMENT OF CHANGE—PARAGRAPH L				
1. Written procedures for MOC including authorization, tracking, and documentation of final resolution	Y	Y	Y	Y
2. Documentation of training of employees on changes and their impact, updating of process safety information and operating procedures	Y	Y	Y	
3. Other:				
INCIDENT INVESTIGATION—PARAGRAPH M				
1. Procedures [written] for performing incident investigations and other requirements of 29 CFR 1910.119, paragraph [o]	Y	Y	Y	Y
2. Assignment of incident investigation team member for any incident involving contractor employees	Y	Y	Y	
3. Other:				
EMERGENCY PLANNING AND RESPONSE—PARAGRAPH N				
1. Emergency action plans in place	Y	Y	O	Y
2. Appropriate emergency actions and information regarding site safety issues transmitted to the contractor	Y	Y		Y

Description of PSM Compliance Issues for Capital Projects under 29 CFR 1910.119	PSM Requirement	Project Requirement	EPC Supply	Client Supply
3. Documentation of training of contractor employees on safety issues and emergency action plans	Y	Y	Y	
4. Additional OSHA training for contractor personnel as required, particularly under 29 CFR 1910.120	Y	Y	Y	
5. Other:				
COMPLIANCE AUDITS—PARAGRAPH O				
1. Compliance audit procedure in place	Y			Y
2. Certification that the facility has been evaluated for compliance with 29 CFR 1910.119	Y			Y
3. Other:				
TRADE SECRETS—PARAGRAPH P				
1. Written policy for dealing with trade secrets		Y	Y	Y
2. Completed confidentiality agreements		Y	J	J
3. Other:				
Project Notes and Clarifications:				

Legend: Y - Yes; N - No; J - Joint Responsibility; A - Approval Only; O - Optional

3.4. REFERENCES

3.4.1. *Specific References*

29 CFR 1910.119. *Process Safety Management of Highly Hazardous Chemicals.* Occupational Safety and Health Administration (OSHA).

29 CFR 1910.120. *Hazardous Waste Operations and Emergency Response.* Occupational Safety and Health Administration (OSHA).

29 CFR 1910.1200. *Hazard Communication.* Occupational Safety and Health Administration (OSHA).

OSHA Instruction CPL 2-2.45A CH-1. September 13, 1994. OSHA Directorate of Compliance Programs.

Responsible Care, Process Safety Code of Management Practices. 1990. Chemical Manufacturers Association (CMA), Washington, D. C.

Kerridge and Vervalin, editors. 1986. *Engineering & Construction Project Management.* Gulf Publishing Company, Houston, Texas.

Guidelines for Hazard Evaluation Procedures, 2nd Edition with Worked Examples. 1992. AIChE-CCPS, New York.

3.4.2 Suggested Reading

40 CFR Part 68. *Risk Management Programs for Chemical Accident Release Prevention.* Environmental Protection Agency (EPA).

4

SUBCONTRACTOR RELATIONSHIPS

This chapter introduces the relationship(s) between the general contractor, his subcontractor(s), and the client. The issue of a seamless contractor relationship under which the subcontractor appears to be one with the general contractor is an important aspect of the relationship. This is the relationship that most clients wish to see, but it is very difficult to implement for various reasons which are discussed. Hopefully, a broader knowledge of this issue by all parties will help the clients understand and deal with the difficulties that this issue presents to the general contractor.

Subcontractor A person or company that by a secondary contract assumes some or all of the obligations of the primary (or general) contractor.

In *Guidelines for Technical Management of Chemical Process Safety*, CCPS states (p. 70) that plant owners should encourage customers, suppliers, and others to adopt the same high standards of risk management as the owner adopts. This advice is valid not only for the owner (client), but also for the general contractor and his subcontractors. Remember, the general contractor may only be considered as good as his weakest subcontractor.

4.1. INTRODUCTION

Understanding and dealing with subcontractor safety issues has become a complex issue over the years, exacerbated by regulatory and legal pressures. General contractors or client companies have found that OSHA will often provide duplicate citations and fines for violations when those violations are committed by subcontractors and their workers. On the legal side, third party lawsuits by subcontractor employees have become a problem for both the general contractor and the client (CII, 1991).

In dealing with regulatory issues, the general contractor must take to heart the roles defined in regulations such as the contractor element of PSM (29 CFR 1910.119, paragraph h). The client's responsibilities will normally include obtaining and evaluating information regarding the contract employer's safety performance and programs, and informing contract employers of the known hazards related to the contractor's work and the process, but these activities may have to be assumed by the general contractor when dealing with subcontractors. Similarly, the general contractor may have to include subcontractors when training employees on applicable provisions of the emergency action plan or safe work practices. The general contractor also should periodically evaluate the performance of subcontractors in fulfilling their obligations under PSM. The subcontractor must then be bound by the contract employer's responsibilities under PSM in relationships with both the general contractor and the client company.

To deal with the legal issues, one should look to Chapter 5 for an introduction to managing risk and discuss pertinent issues with legal staff. It is safe to say that whenever legal complications arise, a clearly stated contractual relationship can minimize the sum of legal fees and allow risks to be apportioned on a reasonable basis.

4.2. SEAMLESS INTERFACES

When setting up contractual arrangements for any project, the client prepares a contract with the general contractor. The general contractor then firms the relationship with all subcontractors, who may range from sole proprietorships to major corporations, depending on the services to be performed. Despite the myriad types of companies and/or persons involved, the client wishes (as much as possible) to consider each to be an employee of the general contractor. The problem becomes even more complex on a small project with multiple services, because micro management of contract labor cannot be budgeted.

In order to provide the client with a *seamless interface* among the various contractors, the general contractor must take the initiative. A seamless interface is one where the point of interaction between client and contractor(s) does not reflect (or need to reflect) the fact that there are many contractors involved. To do so, general contractors must put themselves in their client's position when dealing with subcontractors. In so doing, the general contractor's relationship with each subcontractor will then clearly address safety. In other terms, the general contractor should be a good steward of the client's safe work environment.

In *Managing Subcontractor Safety*, the Construction Industry Institute (CII) Safety Task Force has identified ". . . clear evidence of the role that general contractors play in influencing subcontractor safety. The bottom line is that general contractors play a stronger role in the safety of their subcontractors than do the subcontractors themselves." (1991, p. v) While the CII is focused on construction, experience indicates that the same issues exist for maintenance and other services as well. For example, a sole proprietor who performs leadership on a process hazard analysis must also have a sound relationship with both the client and the general contractor, but should include a firm project scope against which all parties can perform and be measured.

The CII has determined, based on study data, that several factors under the control of the general contractor influence subcontractor safety on small projects, including: (1991, p. 4)

- Effective project management
- Effective job coordination
- Managerial emphasis on job safety
- Strong interpersonal skills of supervisors
- Safe work environment.

It can be suggested that the above factors represent a well-managed project of any sort, whether viewed from a safety perspective or from an overall project completion perspective. This is probably correct, in that the first four of five factors require good management effort and skills on the part of the general contractor. The last factor cited above, a safe work environment, must regularly involve the client as well as the general contractor and subcontractors in order to be properly implemented and monitored. Additional factors include safe work procedures, training, communication, and monitoring of job performance.

In viewing subcontractor safety on larger projects,

> The [CII] task force discovered another factor that significantly impacted the safety performances of the subcontractors. This finding related to the degree or quality of coordination provided by the project management team. In short, the safer projects were those in which the coordination efforts of the project management team significantly assisted the subcontractors in the execution of their work tasks. Perhaps the most clear finding in this regard was that the subcontractors with the better safety records were those who also rated the project management team better on the demonstrated ability to coordinate job activities. (1991, p.8)

While unstated, it is very likely that the "better" project management team will recognize safety as an important aspect, and require contractor and subcontractor safety programs and safety program management as part of the overall project.

Accepting that better project management results in better project execution, safer project implementation, lower costs and a safer process, where does the subcontractor fit in? The answer will depend on the level of participation of the subcontractor in the overall project. For instance, a construction subcontractor on a major EPC project should be directly involved with the client early in the project. On the same project, a fire protection engineer or PSM consultant may only need to attend a single meeting to hear the client's philosophy. Additionally, the client should never be surprised by the appearance of a subcontractor (especially one without proper safety training) when the general contractor is expected.

Depending on the breadth of their scope of work, subcontractors can be included as part of the project management team. At a minimum, subcontractors should periodically attend project management team meetings. Too often, the subcontractor performs under the same contractual terms and conditions that the general contractor established with the client early in the project before many project safety decisions were made. Since the subcontractor relationship (at least contractually) is solely with the general contractor, the subcontractor's contract may not reflect (or be updated to reflect) project decisions relating to safety. Such project decisions or agreements should be passed along to the appropriate subcontractors, but in practice this is difficult to carry out. Exposure to the project management team, including open discussions with the client and other subcontractors, then becomes a necessity in order for the subcontractor to understand and contribute to project safety plans. In short, a seamless interface may only be possible by recognizing that multiple parties do exist, and taking steps to assure that open communication and cooperation exists among the various entities involved. This is team building, and requires trust, openness, flexibility, and sometimes delegation of authority to third parties.

4.3. SHARED LIABILITIES

The CII study *Managing Subcontractor Safety* concluded that owners, general contractors, and subcontractors all have a role in providing a safe work environment. The study also concluded that safer projects resulted in reduced costs to all parties. This means that all companies may be more profitable and (or because) their employees have fewer job-related injuries.

Sharing of responsibilities and, possibly more important, the acceptance of properly apportioned liabilities are important. These issues must be clearly addressed in order to provide the teamwork necessary and the guidelines for communication appropriate to having a successful team effort. In a study by its Contracting Phase II Task Force (1993, p. 27), the CII found that savings can be realized on projects where the parties openly discuss unforeseen risks and the way the cost implications of the risks are to be shared by the parties. Thus, open conversation regarding allocation of risks has been shown to have a cost benefit in the construction industry.

Risk can be defined to encompass several issues. Risks of concern may include business or business interruption risks, property loss, chemical exposure, environmental, or community relations. Managing client–contractor (and subcontractor) risk is the topic of Chapter 5, but underlying issues of trust, accountability, and acceptance of liabilities, where appropriate, must be addressed.

Areas where client–contractor–subcontractor relations might seriously impact process safety, and where open communication and discussion of potential risks will be helpful, include:

- Defining appropriate codes and standards,
- Defining accepted good engineering practices,
- Defining and committing to inherently safer designs,
- Implementing and documenting training of employees,
- Responding to emergency situations,
- Defining inspection requirements and rework policies,
- Authorizing and negotiating project change orders,
- Performing safety reviews for and controlling the implementation of project changes,
- Performing constructability reviews,
- Implementing value engineering efforts.

In plant locations where process safety is of the utmost concern, it is often difficult to determine whether a person is a company employee, contractor employee, or contract employee by observation alone. Such situations are becoming more common as the process safety culture evolves. Sharing of responsibilities, individual accountability, and an understanding of each party's obligations are central to establishing the proper process safety culture.

4.4. DIRECT CONTRACT EMPLOYEES

Many companies, when faced with spot manpower needs, will contract directly with experienced persons or small contractors who may be categorized as direct contract employees. These persons may be required to provide those services that the plant needs, but does not have the staff to perform. Common examples include:

- Calculation of safety relief devices,
- Leadership of process hazard analyses (PHAs),
- Staffing of PHAs,
- Mitigating PHA issues,
- Development of operating procedures,
- P&ID walkdowns/updates,
- Mechanical or maintenance activities,
- Fugitive emissions monitoring,
- Installing distributed control systems, or
- Provision of training programs.

Depending on circumstances, direct contract employees may be contracted directly to the client or subcontracted directly to another contractor. Sometimes, these persons may be former employees of the client company, but in all cases are hired because of a unique expertise which meets a specific need of the client company. Whatever the purpose for their relationship, these employees typically appear to be direct, full benefit employees of the client company. Unlike full benefit employees, however, these persons are probably not covered under any company insurance policies, have no other company benefits, and as a result represent a potential liability for the employer. At a minimum, the employer should ask contract employees to provide protection to the employer comparable to workmen's compensation in the event of injury to the worker. Several alternatives exist for addressing this issue from both the employer and employee viewpoints. Other issues, such as employee supervision, should be addressed contractually to the satisfaction of all parties.

4.5. REFERENCES

4.5.1. *Specific References*

Guidelines for Technical Management of Chemical Process Safety. 1989. AIChE-CCPS, New York.

API RP 2220, CMA Manager's Guide. 1991. *Improving Owner and Contractor Safety Performance.* 1st ed. American Petroleum Institute, and Chemical Manufacturer's Association, Washington, D. C.

29 CFR 1910.119. *Process Safety Management of Highly Hazardous Chemicals*. Occupational Safety and Health Administration (OSHA).

Managing Subcontractor Safety. 1991. Construction Industry Institute. Austin, Texas.

Cost-Trust Relationship. 1993. Construction Industry Institute. Austin, Texas.

Contract Risk Allocation and Cost Effectiveness. 1988. Construction Industry Institute. Austin, Texas.

Responsible Care, Process Safety Code of Management Practices. 1990. Chemical Manufacturers Association (CMA), Washington, D. C.

Recommended Guidelines for Contractor Safety and Health. 1990. Texas Chemical Council Occupational Safety Committee Report.

Improving Construction Safety Performance: A Construction Industry Cost Effectiveness Project Report. 1982. The Business Roundtable, New York.

AGC Guide for a Basic Safety Program. 1990. Associated General Contractors of America, Washington, D.C.

4.5.2 Suggested Reading

API RP 2221, CMA Manager's Guide. 1995. *Implementing a Contractor Safety and Health Program.* 1st ed. American Petroleum Institute, and Chemical Manufacturer's Association, Washington, D. C.

5

MANAGING CLIENT–CONTRACTOR RISK

A major engineering firm was awarded a risk assessment project by a large U.S. city. The contract was worth nearly $1 million, but was contingent on acceptance by the engineering firm of the city's contractual language that required the engineering firm to indemnify the city for any and all liability. Since the project required risk assessments of hazardous chemical processes, and since the worst-case accident scenario could be severe, the engineering firm offered to indemnify the city for an amount not to exceed $3 million.

The city's legal staff was unprepared to deal with contingencies such as those presented by the proposed contractual limits of liability and refused to accept the engineering company's offer. The contract was then awarded to a much smaller firm which accepted the unlimited liability, but had a net worth of less than $500,000. The city thus got its contract language, but not the desired protection.

5.1. INTRODUCTION

In the case of the risk assessment contract above, several important aspects regarding the contractual basis of the client–contractor relationship were not successfully addressed. While the city required that it be indemnified by its contractor, it did not ascertain whether the indemnification was valid. While the contract required the contractor to indemnify the city against consequential damages, the city was not protected due to the limited net worth of the contracting firm. In this chapter, the discussion will encompass many aspects of the legal basis for the client–contractor relationship, and focus on the issue of assigning risks on an equitable basis to all parties. In so doing, the risks will not be out of proportion to the rewards for either party.

In recent years, the cost impact of industrial accidents has grown considerably due in part to the increase in number and size of lawsuits and resulting judgments. To establish the necessary basis for discussion of legal issues, the following definitions must be established:

Indemnity A protection or insurance against loss, damage, or liability. A simple example of indemnification might be a small contractor that indemnifies (releases from responsibility) a client in the event of personal injury to the contractor itself while at the client's facility. A more onerous example would be a general contractor who indemnifies a chemical company and is involved in an on-site accident resulting in multiple deaths or large property and production losses. Such an incident could result in a multitude of lawsuits for which the contractor would be fully responsible.

Consequential Damages Damages incurred as an indirect result of an action. An example of consequential damages might be the loss of production at a process facility when a civil contractor cuts through power lines during construction activities at an adjacent property.

Discovery Discovery is the pretrial procedure of taking depositions or other means to compel disclosure of pertinent factual information. An example of discovery is the issuance of subpoenas for specified records and documents.

Information sharing and data access by all parties has become more common in recent years due to the recognition that a team philosophy is necessary for a successful project. Such regulatory drivers as OSHA PSM, the hazardous waste operations and emergency response standard 29 CFR 1910.120, the hazard communication regulation 29 CFR 1910.1200, and the EPA's right-to-know standards have also opened the information flow. These developments have improved the possibilities for managing risk, but have also in some cases caused parties to retrench contractually. What is considered a fair assignment and acceptance of risk may depend on the parties. Unless risk allocation is truly fair, the separate parties will be defensive in order to guard their individual interests.

While risk management of business issues is generally consistent with process risk management, there is a difference between the two. In both cases, the first steps include:

- Identification of the type of risk that might be encountered,
- Attempts to eliminate the risk,
- Attempts to reduce the risk.

Additionally, for business risks a final option is to transfer the risk to another party such as an insurance carrier, the client, the contractor, or the subcontractor. There is probably a parallel in some areas of process risk management, such as the transfer of risk to a third party by having that party generate a chemical that must be manufactured under hazardous operating conditions, but generally the transferral option is for business issues only.

Unfortunately, process safety issues have not been afforded much attention by industry personnel who normally deal with risk management issues. Process safety generally has been categorized under the headings of technical or loss prevention issues. The recognition that process safety is more comprehensive, and the fact that it varies according to the specific chemical process or specific site conditions must be recognized by project and risk managers.

There are many unknowns with the legal issues surrounding implementation of PSM programs. To date, very little case law exists to define or clarify the issues. Unlike specification-based regulations, the performance-based PSM standard is less clear on its requirements and is subject to debate. For instance, what is accepted good engineering practice for one or two firms (or industries) may be unacceptable to others due to their unique operating histories. OSHA historically has audited companies against defined standards, not "good practices," which are sometimes undefinable. For these reasons, PSM issues should be closely attended to during the development of the contractor–client relationship.

There is a last issue to be addressed, and that is the need for true authority to make the risk management decisions. Too often, contract meetings are held without the appropriate personnel or information to make informed decisions. As a result, contracting parties will seek to avoid risk completely (by transferring the risk to the other party). Lacking legal precedent, standard contract guidelines also may not exist. Owners and contractors then find it difficult to agree on risk-sharing formulas that are perceived as equitable for all parties. A reasonable approach would be for the project managers to include legal and insurance representatives in the contract discussions. Otherwise, once decisions have been made and either party determines that its position is unfair, the relationship breaks down unless all parties are open to reasonable discussion or mediation.

5.2. INDEMNIFICATION

> Owners should take particular note of the magnitude of third party liability costs as one of the indirect costs of accidents. Litigation against a third party has become more common in recent years, and dollar losses in some jurisdictions can be significant for the owner when an employee of a contractor sustains an injury or illness. Agreements of indemnification (hold harmless clauses) sometimes tend to be ineffectual in protecting owners from either dollar loss or adverse publicity. However, when carefully drafted, such clauses can provide significant protection to owners and should be considered in all contracts. (Business Roundtable, p. 16)

The above quotation provides valuable insight into the need for indemnification. The approach is slanted in the direction of the owners (clients), however. As established in Section 4.3, the allocation of appropriate risks among all parties must be considered.

In *Contract Risk Allocation and Cost Effectiveness*, the CII (1988) studied four particular contract clauses—indemnity, consequential damages, differing conditions, and delays. The stated goal was to improve cost effectiveness and participant relationships on projects through risk allocation and equity on construction projects. Not surprisingly, findings showed that inequities in the four clauses had a negative impact on the contractor–client relationship. Allocation of risks with clear statements of responsibilities can avoid this problem. For instance, a contract that states that the client will specify and install the initial lubricants on rotating equipment during pre-commissioning leaves no possibility for confusion about responsibility if a bearing seizes due to lack of lubrication at startup.

So what is the appropriate allocation of risk between a contractor and the client on an indemnity clause? Again, one must look at the relationship and who will gain most from the relationship. Often, an indemnity insurance policy can be obtained by a contractor on a major project that will protect all parties to the limits of the policy. Some contractors, especially in construction or other manpower-intensive services, cannot operate without the insurance. In these cases, the practical solution might be to have the contractor absorb all liabilities which might be coverable under the policy. Conversely, a small contractor who obtains a one-month consulting contract might have to double his price if the client requires him to have extensive insurance. The small contractor normally warrants its services to the extent of the contract amount, but may balk at accepting any additional liabilities.

5.3. CONSEQUENTIAL DAMAGES

Consequential damages may be the most difficult and tedious of the contractual issues to resolve, because it is more difficult to foresee the possible scenarios involving consequential damages. For example, an incident that occurs at a chemical plant might implicate or involve (directly or indirectly) the engineering designers, the constructors, equipment suppliers, equipment manufacturers, several subcontractors, miscellaneous parties on-site, and the chemical company's own personnel. While each employer might be expected to have its relationship with its own employees well defined, third party lawsuits from every other entity or person are possible.

In the example at the start of this chapter, the engineering firm offered a limit of liability in an attempt to establish an equitable position on consequential damages. The city, however, could not (under the city's regulations) allow the public to contractually accept the liabilities posed by the city's own public works. Imagine, however, the consequences of a catastrophic release of chlorine at a public swimming pool on a warm summer day. This scenario also points to the need for a contract to establish the relative liabilities after the project is completed. For example, should a constructor be held liable once a chemical company has operated, maintained, and occasionally modified equipment over a period of ten years?

5.4. DIFFERING CONDITIONS

Many contractors commonly utilize the following verbiage, or something very similar, in limiting their liability under contracts.

> In the performance of its services, the Contractor will rely upon the accuracy and completeness of the information and data provided by the Client. Specifically excluded from the Contractor's scope of services is any verification or confirmation of information or data provided by the Client.

Differing conditions are conditions that change which cannot be reasonably foreseen. Many such conditions are the direct result of client-supplied documents that may not be entirely factual. For instance, a piping isometric for a piping tie-in that has been supplied to the contractor with improper coordinates would be a differing condition if the contract did not include field verification by the contractor. A more serious example would be the discovery of hazardous soil contaminants during civil work on a lump sum project where there is no contractual remedy. The above clause is designed to address some of these issues. While the clause allows the contractor to avoid liabilities for what would be the client's mistake, he cannot avoid liability for his own misinterpretation of the same documents.

Contractors may assume considerable risks under lump sum contracts, with differing conditions representing only a portion of the risk. To allow for credible contingency planning, project managers need to understand process risks and process risk analyses in addition to other risk management data. Since subcontractors on a lump sum project are typically also on a lump sum basis, the subcontractors might be included in pre-bid meetings if the subcontractors have been identified, thereby allowing the subcontractor to make his own informed risk management decisions.

Consider the example of mitigation following a PHA on a lump sum EPC project. While a PHA is usually part of the project basis, there may be a difference of opinion regarding the resolution of a specific item. For instance, when a client wants an engineered solution for an identified hazard and the engineering company believes that an administrative (procedural) fix adequately represents "good engineering practice," who will pay for the engineered solution? Remember, this might impact engineering, procurement, construction, and subcontractors—many of whom can be essentially finished with their contractual obligations by the time the PHA mitigation is to be implemented.

5.5. DELAYS

Projects involving complex design efforts, extensive procurement, and/or major construction are inevitably scheduled optimistically. This is often thrust on the contractor by direct bid requirement (such as thirty months for mechanical completion of a grass roots facility), with penalties for missing the schedule date. To limit its exposure, the engineering contractor firm will typically price its efforts on the worst-case basis, with the hope of meeting the milestones to maximize profits. One can see that the schedule is now likely to be fictitious and counterproductive for the client.

Delays can occur for many reasons; a sampling includes:

- Delays in client decision making,
- Changes in plant feedstocks,
- Loss of key project personnel,
- Long delivery periods for critical equipment,
- Deficiencies of delivered equipment,
- Labor disputes,
- Weather problems.

Interestingly, most of the above causes for delays cannot be controlled by the contracting parties. A contractual basis that addresses any delays that might be foreseen and sets in place a procedure to mediate any losses due to delays will allow the parties to focus more on the project needs than any contractual liabilities.

5.6. ATTORNEY–CLIENT PRIVILEGE

Due to the probability that internal company documents will become discoverable, or open to review in the event of a lawsuit being filed, the issue of attorney–client privilege should be understood by all parties. Most companies find that the use of outside attorneys is a necessity to keep information privileged until it is deemed appropriate to disseminate (such as during a lawsuit). While OSHA PSM requires open and timely incident investigations, further investigations, which are sometimes more detailed and highly technical, can also occur. Examples include forensic investigations, detailed dispersion modeling or blast calculations, or toxicological studies. In such cases, the outside attorneys must hire outside consultants who are not available to the client directly in order to maintain attorney–client privilege. If the consultant is hired directly by the company, he may be deemed a party to the routine investigation and his work may become discoverable while in progress (Markham and Clark).

Another aspect of discoverability is that the outside team should only discuss findings with the company's lawyers, avoiding discussions with other company personnel. Written documents should also be minimized, because memos to files or to other employees besides lawyers may be discoverable. This issue of privilege is one that will come under more scrutiny soon, as industry and OSHA address the regulatory requirements under PSM for open access to information. Already, issues have been raised surrounding the requirement of 29 CFR 1910.119, paragraph m, incident investigation, to establish a system to promptly address and resolve the incident report findings and recommendations, document the resolutions and corrective actions, and review the findings with all affected personnel. Unless proper care is taken, the companies involved might infringe upon an individual's rights to privacy when providing pertinent information to other employees.

A last word of caution, when dealing within the regulatory environment, is that any person in a management or supervisory position may face criminal charges in the event that a known, potentially serious safety condition goes unreported (Markham and Clark).

5.7. TRUST

The Contracting Phase II Task Force of the CII prepared a booklet titled *Cost–Trust Relationship* where it defined trust as "the confidence and reliance one party has in the professional competence and integrity of the other party to successfully execute a project in the spirit of open communication and fairness." The task force further summarized trust in the following perceptions: (CII, 1993, p. 2)

- A belief that both parties will do what they have said they will do.
- A willingness for both parties to risk being vulnerable to the other, supported by the belief that neither party will take advantage of the situation.
- Sensitivity and active dedication to each other's needs.
- Candid communication about how each sees the relationship.

The AIChE code of professional ethics offers a sound basis for dealing with contractor–client relationships and managing the risks of those relationships. The code states that members

> shall uphold and advance the integrity, honor and dignity of the engineering profession by: being honest and impartial and serving with fidelity their employers, their clients, and the public; striving to increase the competence and prestige of the engineering profession; and using their knowledge and skill for the enhancement of human welfare. To achieve these goals, members shall:
>
> - Hold paramount the safety, health, and welfare of the public in performance of their professional duties.
> - Formally advise their employers or clients (and consider further disclosure, if warranted) if they perceive that a consequence of their duties will adversely affect the present or future health or safety of their colleagues or the public.
> - Accept responsibility for their actions and recognize the contributions of others; seek critical review of their work and offer objective criticism of the work of others.
> - Issue statements or present information only in an objective and truthful manner.
> - Act in professional matters for each employer or client as faithful agents or trustees, and avoid conflicts of interest.
> - Treat fairly all colleagues and co-workers, recognizing their unique contributions and capabilities.
> - Perform professional services only in areas of their competence.
> - Build professional reputations on the merits of their services.
> - Continue their professional development throughout their careers, and provide opportunities for the professional development of those under their supervision.

Simply put, *a relationship founded upon the principles of professional ethics and integrity, supported by a history of performance, will provide the necessary environment to effectively manage client-contractor risks.*

5.8. REFERENCES

5.8.1. Specific References

API RP 2220, CMA Manager's Guide. 1991. *Improving Owner and Contractor safety Performance.* 1st ed. American Petroleum Institute, and Chemical Manufacturer's Association, Washington, D. C.

29 CFR 1910.119. *Process Safety Management of Highly Hazardous Chemicals.* Occupational Safety and Health Administration (OSHA).

Managing Subcontractor Safety. 1991. Construction Industry Institute. Austin, Texas.

Contract Risk Allocation and Cost Effectiveness. 1988. Construction Industry Institute. Austin, Texas.

Cost-Trust Relationship. 1993. Construction Industry Institute. Austin, Texas.

Responsible Care, Process Safety Code of Management Practices. 1990. Chemical Manufacturers Association (CMA), Washington, D. C.

Recommended Guidelines for Contractor Safety and Health. 1990. Texas Chemical Council Occupational Safety Committee Report.

Improving Construction Safety Performance: A Construction Industry Cost Effectiveness Project Report. 1982. The Business Roundtable, New York.

Markham, G., and Clark, C., *Accident Investigation Under OSHA PSM: A Contractor's View*, 1993. Presented to American Contractors Insurance Group Safety/Claims Workshop, Baltimore, Maryland.

Wideman, R. M., ed., *Project and Program Risk Management: A Guide to Managing Project Risks and Opportunities.* 1992. Project Management Institute. Upper Darby, Pennsylvania.

5.8.2 Suggested Reading

Allocation of Insurance-Related Risks and Costs on Construction Projects. 1993. Construction Industry Institute. Austin, Texas.

Derk, Walter D., *Insurance for Contractors,* 4th ed. 1974. James Risk Management. Chicago, Illinois.

APPENDIX A

CONTRACTOR CHECKLISTS

On following pages, three separate contractor checklists are presented as examples of how a company might approach the issue of preselecting, qualifying, and monitoring contractors to perform activities subject to U.S. Department of Labor Occupational Safety and Health Administration (OSHA) regulations, including the Process Safety Management (PSM) regulation 29 CFR 1910.119. The first of the three documents is intended for use in the initial contractor evaluation and qualification, the second addresses prequalification and selection, and the third addresses and documents contractor training and information dissemination.

These checklists each have a separate role in assisting the plant owner in compliance with the contractors element of OSHA PSM, but also serve as a basis for selecting, qualifying, and monitoring contractors in general.

PROCESS SAFETY MANAGEMENT
Contractor Qualifications Evaluation Checklist

To Be Completed by the Contract Originator/Administrator

NOTE: Information should be provided to the Safety Manager to ensure that the Contractor meets Company requirements.

A. **Contractor Information**

1. Name: ______________________________
2. Contact Name: ______________________________
3. Phone Number: ______________________________
4. Mailing Address (City, State, Zip Code):
____
____

B. **Safety**

Ensure that the Contractor has provided a copy of the following information:

	Yes	No
1. A written safety policy endorsed and supported by management.		
2. A safety manual copy		
3. Copies of OSHA 200 forms for the last three years with the total number of employees and/or total hours for the time period on the 200 forms		
4. Worker's Compensation Experience Modification Rates (EMR) where applicable		
5. A written narrative of how the Contractor will comply with the provisions of OSHA's 29 CFR, Part 1910.119, Section (h) (3), Contract Employer Responsibilities		
6. A description of safety, health, and fire training that Contractor's employees have received or will receive		

PROCESS SAFETY MANAGEMENT
Contractor Qualifications Evaluation Checklist

SIGNATURE:

______________________________ ______________

Contract Originator **Date**

Is Contractor qualified: Yes ☐ No ☐

If Contractor is not qualified, please explain: ______________________________

____

____

____

SIGNATURE:

______________________________ ______________

Safety Manager **Date**

PROCESS SAFETY MANAGEMENT
Contractor Prequalification and Selection

Contractor: ______________________ Date: ____________

Contractor Superintendent: ______________________

Company Location: ______________________

Project: ______________________

Company Representative: ______________________

No.	Item	Yes	No	Not Applicable
Training				
1	Does the contractor ensure that all of its personnel are competent to perform the work required?			
2	Does the contractor ensure that its newly assigned workers are directly supervised by a competent worker?			
3	Does the contractor ensure that jobs requiring certification are performed by workers who possess the appropriate documentation and certificates?			
4	Does the contractor provide training for its managers and supervisors to ensure that they are capable of administering the safety program?			
5	Is there a training plan for contractor employees?			
6	Does the training plan address:			
	Safety and health?			
	Material safety data sheets and hazard communication program?			
	Safety orientation?			
	First aid and cardiopulmonary resuscitation?			
	Fire fighting?			
	Firewatch?			
	Hydrogen sulfide?			
	Transportation of hazardous materials?			
	Fall protection?			
	Forklift and crane operations?			

PROCESS SAFETY MANAGEMENT
Contractor Prequalification and Selection

No.	Item	Yes	No	Not Applicable
Training (Continued)				
	Housekeeping?			
	Entry into confined spaces and requirements for standby personnel?			
	Permit systems?			
	Abrasive blasting and hydroblasting?			
	Respiratory protection?			
	Use of personal protective equipment?			
	Control of hazardous energy sources?			
	Excavating, shoring, and trenching?			
	Emergency response plan?			
7	Is there documentation on file to verify that the training has been completed?			
8	Does the documentation include employees' names, course dates, instructors' names, the length of the courses, and outlines or descriptions of the course content?			
9	Does the contractor require retraining or refresher training?			
10	Does the contractor expect the facility company to provide assistance with the training?			
Incident Investigation and Analysis				
11	Are all incidents investigated to determine their cause, and is corrective action taken?			
12	Are serious incidents and near misses reported to the company immediately?			
13	Does the contractor have written incident investigation procedures?			
14	Is a report completed for:			
	Facilities?			
	OSHA-recordable accidents?			
	Vehicle accidents?			

PROCESS SAFETY MANAGEMENT
Contractor Prequalification and Selection

No.	Item	Yes	No	Not Applicable
Incident Investigation and Analysis (Continued)				
	Equipment damage?			
	Spills?			
	Fires?			
	Near misses?			
	Contractor injury resulting from company action or equipment?			
15	After an incident, are required reports submitted to the company within 48 hours?			
16	Does a safety committee assist in the investigation and follow-up?			
17	Do the contractor's supervisors or managers ensure that, as a result of the investigation, corrective action is taken?			
18	Does the contractor develop a monthly statistical summary that illustrates its safety performance?			
19	Does the company receive a copy of this summary and photocopies of the OSHA Form 200 logs?			
Safety Responsibility				
20	Has the contractor assigned a competent person as the safety representative?			
21	Does the safety representative have sufficient authority to implement change?			
22	Is the contractor's management aware of its role in an effective safety program?			
23	Are the contractor's employees aware of their roles in an effective safety program?			
24	Does the contractor have a disciplinary action program that includes safety and health issues, and is the program enforced?			
25	Does the contractor have a safety policy statement endorsed by its top management?			
Safety Reviews				
26	Does the contractor conduct ongoing safety inspections?			

PROCESS SAFETY MANAGEMENT
Contractor Prequalification and Selection

No.	Item	Yes	No	Not Applicable
Safety Reviews (Continued)				
27	Does the contractor have a hazard identification program that allows employees to report unsafe acts or conditions?			
28	Are inspection records kept on file, and are they available for review by the company?			
29	Are concerns reviewed, and are corrective actions taken?			
30	Is there written evidence that concerns have been reviewed and corrective actions have been taken?			
31	Is there a system to notify the company of safety and health programs that are not created by the contractor but could impact personnel at the site?			
Emergency Response				
32	Are the contractor's employees aware of their role in an emergency?			
33	Have exit routes and meeting areas where head counts are to be performed been established?			
34	Do all employees receive emergency response training?			
35	Does the contractor know how to report an emergency?			
36	Have instructions been given to check wind conditions in the event of a fire or gas release and to evacuate upwind or crosswind?			
37	Are emergency telephone numbers posted throughout the site?			
38	Have the contractor's employees received specific instructions about vehicle use during and after an emergency?			
39	Is the contractor involved in facility drills?			
Job Analysis and Observation				
40	Are all critical jobs identified and analyzed?			
41	Are the procedures for critical jobs written and reviewed with the contract employees before the work begins?			
Permit Systems				
42	Does the contractor ensure that all work permit systems are followed?			

PROCESS SAFETY MANAGEMENT
Contractor Prequalification and Selection

No.	Item	Yes	No	Not Applicable
Permit Systems (Continued)				
43	Does the contractor conduct audits to verify that all work permit systems are followed, and is disciplinary action implemented in the event of noncompliance?			
44	Is there a method that allows the contractor to provide the company with feedback about the effectiveness of the permit system and the impact of the process on the performance of the work and on the contract?			
First Aid and Medical Care				
45	Does the contractor have personnel trained to administer first aid and cardiopulmonary resuscitation?			
46	Are adequate first aid supplies available on site?			
47	Have the first aid kits been approved by a physician?			
48	Have arrangements been made with the company, ambulance service, hospital, or others to handle medical care ranging from first aid to life-threatening injuries and illnesses?			
49	Does the contractor utilize a "physician under standing orders"?			
50	Does the contractor distribute prescription medications?			
Personal Protective Equipment				
51	Are specific rules developed and communicated concerning the proper use of personal protective equipment?			
52	Is the necessary personal protective equipment available on site?			
53	Has a maintenance program been established to assure that personal protective equipment is maintained in satisfactory condition?			
54	Is the use of personal protective equipment enforced by the contractor's management?			
Communications and Meetings				
55	Before a job begins, does the contractor hold a meeting to address safety issues?			
56	Do the contractor's supervisors mention safety as part of every work assignment?			

PROCESS SAFETY MANAGEMENT
Contractor Prequalification and Selection

No.	Item	Yes	No	Not Applicable
Communications and Meetings (Continued)				
57	Are tailgate meetings held? If they are, how long do they last, how often are they held, and how effective are they?			
58	Are the topics discussed during the meetings, names of those who attended, dates, and any comments or concerns documented?			
59	Are problems identified during the tailgate meetings acted on by the contractor's management?			
60	Does the contractor make use of safety committees?			
61	Does the contractor's management actively demonstrate support for improved safety programs?			
62	Does the contractor's management hold periodic meetings with the company's management specifically to discuss safety, health, and job performance?			
63	Are the contractor's employees able to comminicate all safety and health problems to their management?			
64	Is a meeting held with the contractor at the end of the project to review safety performance?			

PROCESS SAFETY MANAGEMENT
Contractor Information and Training

Contract Employer Training for Local Facility

No.	Item	Yes	No	Not Applicable
1	Did the contractor supply a list of all contract employees working in the process facility?			
2	Does the contractor understand the fire hazards in this facility?			
3	Does the contractor understand the explosion hazards in this facility?			
4	Does the contractor understand the toxic release hazards in this facility?			
5	Does the contractor know the exit routes to take in the event of:			
	One of the above emergencies?			
	Wind direction indicators?			
6	Does the contractor know where to meet for a head count after exiting the facility during one of the above mentioned emergencies?			
7	Does the contractor understand its role:			
	In an emergency?			
	Reporting procedures?			
	Emergency phone numbers?			
8	Does the contractor understand that contract employees will be involved in the facility drills?			
9	Does the contractor understand the work authorization permit system:			
	Hot work?			
	Confined space entry?			
	Excavation?			
	Control of hazardous energy sources?			
10	Does the contractor understand the Entry/Exit control system?			

Contractor has received instructions and training for the above elements.

Contractor: ______________________________ Date: ____________

Contractor Superintendent: ______________________________

Company Location: ______________________________

Project: ______________________________

Company Supervisor Verifying Training: ______________________________

APPENDIX B

STANDARDIZED PREQUALIFICATION FORM (PQF)

A standardized contractor prequalification form (PQF) is presented on the following pages. This form was incorporated in Appendix A of API Recommended Practice 2221, CMA Manager's Guide, *Implementing a Contractor Safety and Health Program*. The form has been endorsed by the Texas Chemical Council, and a software copy of the form is available for a nominal fee from the Houston Business Roundtable, 8031 Airport Blvd., Suite 118, Houston, Texas, 77061-4177. For the API RP 2221 document, please contact either the American Petroleum Institute, or Chemical Manufacturer's Association, both of whom are headquartered in Washington, D. C.

GENERAL INFORMATION		
1. Company Name:	Telephone:	Fax:
Street Address:	Mailing Address:	
2. Officers	Years With Company	
President:		
Vice President:		
Treasurer:		
3. How many years has your organization been in business under your present firm name?		
4. Parent Company Name:		
City:	State:	Zip:
Subsidiaries:		
5. Under Current Management Since (Date):		
6. Contact for Insurance Information:		
Title:	Telephone:	Fax:
7. Insurance Carrier(s):		
Name	Type of Coverage	Telephone
8. Are you self insured for Worker's Compensation Insurance?	Yes ☐	No ☐
9. Contact for Requesting Bids:		
Title:	Telephone:	Fax:
10. PQF Completed By:		
Title:	Telephone:	Fax:

ORGANIZATION		
11. Form of Business: Sole Owner ☐	Partnership ☐	Corporation ☐
12. Percent Minority/Female Owned:	EEO Category:	

13. Describe Services Performed: SIC Code:

- ☐ Construction
- ☐ Construction Design
- ☐ Original Equipment Manufacturer and Installer
- ☐ Project Maintenance
- ☐ Maintenance
- ☐ Original Equipment Manufacturer and Maintenance
- ☐ Service work (e.g., janitorial, clerical, etc.)
- ☐ Manpower and Resource
- ☐ Other

14. Describe Additional Services Performed:

15. List other types of work within the services you normally perform that you subcontract to others:

16. Attach a list of major equipment (e.g., cranes, JLGs, forklifts) your company has available for work at this facility and the method of establishing competency to operate.

17. Do you normally employ? ☐ Union Personnel ☐ Non-Union Personnel

If union, list trades/locals:

18. Company Paid Benefits - Do you have or provide:

a.	Health insurance	Yes	☐	No	☐
b.	Dental insurance	Yes	☐	No	☐
c.	Paid vacation	Yes	☐	No	☐
d.	Paid holidays	Yes	☐	No	☐
e.	Paid sick leave	Yes	☐	No	☐
f.	Educational reimbursement program	Yes	☐	No	☐
g.	Employee profit sharing	Yes	☐	No	☐

19. Annual Dollar Volume for the Past Three Years:	19 ______ $	19 ______ $	19 ______ $

20. Largest Job During the Last 3 Years: $

21. Your Firm's Desired Project Size:	Maximum:	Minimum:

22. D&B Financial Rating:	Annual Sales $	Net Worth: $

COMPANY WORK HISTORY

23. Major jobs in progress:

Customer/Location	Type of Work	Size $M	Customer Contact	Telephone

24. Major jobs completed in the past three years:

Customer/Location	Type of Work	Size $M	Customer Contact	Telephone

25. Are there any judgments, claims or suits pending or outstanding against your company?
If yes, please attach details. Yes ☐ No ☐

26. Are you now or have you ever been involved in any bankruptcy or reorganization proceedings?
If yes, please attach details. Yes ☐ No ☐

SAFETY AND HEALTH PERFORMANCE

27. Workers Compensation Experience Modification Rate (EMR) Data

a. EMR is:

☐ Interstate rate

☐ Intrastate rate

☐ Monopolistic State rate

☐ Dual rate

b. EMR for last three years:

__________ 199___

__________ 199___

__________ 199___

c. State or Origin:

d. EMR Anniversary Date:

28. Injury and Illness Data:

a. Employee hours worked last three years (excluding subcontractors)	Hours / Year	19 ___	19 ___	19 ___
	Field			
	Total			

b. Provide the following data (excluding subcontractor) using your OSHA 200 Forms for the past three (3) years:

	19		19		19	
	No.	Rate	No.	Rate	No.	Rate
Injury related fatality $Rate = \frac{Total\ Col.\ 1 \times 200{,}000}{Total\ Employee\ Hours}$						
Lost workday cases injuries involving days away from work, or days of restricted work activity or both. $Rate = \frac{Total\ Col.\ 2 \times 200{,}000}{Total\ Employee\ Hours}$						
Lost workday case injuries involving days away from work. $Rate = \frac{Total\ Col.\ 3 \times 200{,}000}{Total\ Employee\ Hours}$						
Injuries involving medical treatment only. $Rate = \frac{Total\ Col.\ 6 \times 200{,}000}{Total\ Employee\ Hours}$						

	19		19		19	
	No.	Rate	No.	Rate	No.	Rate
Total OSHA Recordable Injury Rate $Rate = \frac{(Total\ Col.\ 1 + 2 + 6) \times 200{,}000}{Total\ Employee\ Hours}$						
Illness related fatality $Rate = \frac{Total\ Col.\ 8 \times 200{,}000}{Total\ Employee\ Hours}$						
Lost workday case illnesses involving days away from work, or days of restricted work activity, or both. $Rate = \frac{Total\ Col.\ 9 \times 200{,}000}{Total\ Employee\ Hours}$						
Lost workday case illnesses involving days away from work $Rate = \frac{Total\ Col.\ 10 \times 200{,}000}{Total\ Employee\ Hours}$						
Illnesses not involving lost workdays or restricted workdays $Rate = \frac{Total\ Col.\ 13 \times 200{,}000}{Total\ Employee\ Hours}$						
Total OSHA Recordable Illness Rate $Rate = \frac{(Total\ Col.\ 8 + 9 + 13) \times 200{,}000}{Total\ Employee\ Hours}$						
Total OSHA Recordable Injury/Illness Rate $Rate = \frac{(Total\ Col.1+2+6+8+9+13) \times 200{,}000}{Total\ Employee\ Hours}$						

Notes:
1. Data should be the best available data applicable to the work in this region or area.
2. If your company is not required to maintain OSHA 200 forms, please provide information from your Worker's Compensation insurance carrier itemizing all claims for the last three years.

29. Have you received any regulatory (EPA, OSHA, etc.) citations in the last three years?
If yes, please attach copies Yes ☐ No ☐

SAFETY AND HEALTH MANAGEMENT

30. Highest ranking safety/health professional in the company:

Title:	Telephone:	Fax:

31. Do you have or provide:
    a. Full time Safety/Health Director Yes ☐ No ☐
    b. Full time Safety/Health Supervisor Yes ☐ No ☐
    c. Full Time Job Safety/Health Coordinator Yes ☐ No ☐

32.	Do you have or provide:					
	a.	Safety/Health incentive program	Yes	☐	No	☐
	b.	Company paid safety/health training	Yes	☐	No	☐
SAFETY AND HEALTH PROGRAMS AND PROCEDURES						
33.	Do you have a written Safety & Health Program?		Yes	☐	No	☐
	Does the program address the following key elements?					
	•	Management commitment and expectations	Yes	☐	No	☐
	•	Employee participation	Yes	☐	No	☐
	•	Accountabilities and responsibilities for managers, supervisors, and employees	Yes	☐	No	☐
	•	Resources for meeting safety & health requirements	Yes	☐	No	☐
	•	Periodic safety and health performance appraisals for all employees	Yes	☐	No	☐
	•	Hazard recognition and control	Yes	☐	No	☐
34.	Does the program include work practices and procedures such as:					
	a.	Equipment Lockout and Tagout (LOTO)	Yes	☐	No	☐
	b.	Confined Space Entry	Yes	☐	No	☐
	c.	Injury & Illness Recording	Yes	☐	No	☐
	d.	Fall Protection	Yes	☐	No	☐
	e.	Personal Protective Equipment	Yes	☐	No	☐
	f.	Portable Electrical/Power Tools	Yes	☐	No	☐
	g.	Vehicle Safety	Yes	☐	No	☐
	h.	Compressed Gas Cylinders	Yes	☐	No	☐
	i.	Electrical Equipment Grounding Assurance	Yes	☐	No	☐
	j.	Powered Industrial Vehicles (Crane, Forklifts, JLGs, etc.)	Yes	☐	No	☐
	k.	Housekeeping	Yes	☐	No	☐
	l.	Accident/Incident Reporting	Yes	☐	No	☐
	m	Unsafe Condition Reporting	Yes	☐	No	☐
	n.	Emergency Preparedness, including evacuation plan	Yes	☐	No	☐
	o.	Waste Disposal	Yes	☐	No	☐

35. Do you have written programs for the following:

a. Hearing Conservation — Yes ☐ No ☐

b. Respiratory Protection — Yes ☐ No ☐

Where applicable, have employees been:

☐ Trained

☐ Fit tested

☐ Medically approved

c. Hazard Communication — Yes ☐ No ☐

d. Program to support the contractor requirements of the OSHA Process Safety Management of Highly Hazardous Chemicals; Explosives and Blasting Agents Standard (29 CFR 1910). — Yes ☐ No ☐

36. Do you have a substance abuse program? — Yes ☐ No ☐

If yes, does it include the following?

- Pre-placement Testing — Yes ☐ No ☐
- Random Testing — Yes ☐ No ☐
- Testing for Cause — Yes ☐ No ☐
- DOT Testing — Yes ☐ No ☐

37. Do your employees read, write and understand English such that they can perform their job tasks safely without an interpreter? — Yes ☐ No ☐

If no, provide a description of your plan to assure that they can safely perform their jobs.

38. Medical

a. Do you conduct medical examinations for:

- Pre-placement — Yes ☐ No ☐
- Pre-placement Job Capability — Yes ☐ No ☐
- Hearing Function (Audiogram) — Yes ☐ No ☐
- Pulmonary — Yes ☐ No ☐
- Respiratory — Yes ☐ No ☐

b. Describe how you will provide first aid and other medical services for your employees while on site.
Specify who will provide this service: ______________________________

c. Do you have personnel trained to perform first aid & CPR? — Yes ☐ No ☐

39. Do you hold site safety and health meetings for:

Field Supervisors	Yes ☐	No ☐	Frequency_______________
Employees	Yes ☐	No ☐	Frequency_______________
New Hires	Yes ☐	No ☐	Frequency_______________
Subcontractors	Yes ☐	No ☐	Frequency_______________

Are the safety and health meetings documented? — Yes ☐ No ☐

40.	Personal Protection Equipment (PPE)					
	a.	Is applicable PPE provided for employees?	Yes	☐	No	☐
	b.	Do you have a program to assure that PPE is inspected and maintained?	Yes	☐	No	☐
41.	Do you have a corrective action process for addressing individual safety and health performance deficiencies?		Yes	☐	No	☐
42.	Equipment and Materials					
	a.	Do you have a system for establishing applicable health, safety, and environmental specifications for acquisition of materials and equipment?	Yes	☐	No	☐
	b.	Do you conduct inspections on operating equipment (e.g., cranes, forklifts, JLGs) in compliance with regulatory requirements?	Yes	☐	No	☐
	c.	Do you maintain operating equipment in compliance with regulatory requirements?	Yes	☐	No	☐
	d.	Do you maintain the applicable inspection and maintenance certification records for operating equipment?	Yes	☐	No	☐
43.	Subcontractors					
	a.	Do you use safety and health performance criteria in selection of subcontractors?	Yes	☐	No	☐
	b.	Do you evaluate the ability of subcontractors to comply with applicable health and safety requirements as part of the selection process?	Yes	☐	No	☐
	c.	Do your subcontractors have a written Safety & Health Program?	Yes	☐	No	☐
	d.	Do you include your subcontractors in:				
		• Safety & Health Orientation	Yes	☐	No	☐
		• Safety & Health Meeting	Yes	☐	No	☐
		• Inspections	Yes	☐	No	☐
		• Audits	Yes	☐	No	☐
44.	Inspections and Audits					
	a.	Do you conduct safety and health inspections?	Yes	☐	No	☐
	b.	Do you conduct safety and health program audits?	Yes	☐	No	☐
	c.	Are corrections of deficiencies documented?	Yes	☐	No	☐

SAFETY AND HEALTH TRAINING

45. Craft Training

a.	Have employees been trained in appropriate job skills?	Yes	☐	No	☐
b.	Are employees job skills certified where required by regulatory or industry consensus standards?	Yes	☐	No	☐

c. List crafts which have been certified:

____

____

____

46. Safety & Health Orientation

		New Hires				Supervisors			
a.	Do you have a Safety & Health Orientation Program for new hires and newly hired or promoted supervisors?	Yes	☐	No	☐	Yes	☐	No	☐
b.	Does program provide instruction on the following:								
	•New Worker Orientation	Yes	☐	No	☐	Yes	☐	No	☐
	•Safe Work Practices	Yes	☐	No	☐	Yes	☐	No	☐
	•Safety Supervision	Yes	☐	No	☐	Yes	☐	No	☐
	•Toolbox Meetings	Yes	☐	No	☐	Yes	☐	No	☐
	•Emergency Procedures	Yes	☐	No	☐	Yes	☐	No	☐
	•First Aid Procedures	Yes	☐	No	☐	Yes	☐	No	☐
	•Incident Investigation	Yes	☐	No	☐	Yes	☐	No	☐
	•Fire Protection and Prevention	Yes	☐	No	☐	Yes	☐	No	☐
	•Safety Intervention	Yes	☐	No	☐	Yes	☐	No	☐
	•Hazard Communication	Yes	☐	No	☐	Yes	☐	No	☐

c. How long is the orientation program? ______________ Hours

47. Safety & Health Training

a.	Do you know the regulatory safety and health training requirements for your employees?	Yes	☐	No	☐
b.	Have your employees received the required safety and health training and retraining?	Yes	☐	No	☐
c.	Do you have a specific safety and health training program for supervisors?	Yes	☐	No	☐

48. Training Records

a. Do you have safety and health and crafts training records for your employees? Yes ☐ No ☐

b. Do the training records include the following:

Employee identification	Yes ☐	No ☐
Date of training	Yes ☐	No ☐
Name of trainer	Yes ☐	No ☐
Method used to verify understanding	Yes ☐	No ☐

c. How do you verify understanding of training? (Check all that apply.)

☐ Written test ☐ Job Monitoring

☐ Oral test Other (List) ______________________

☐ Performance test ______________________

INFORMATION SUBMITTAL

Please provide copies of checked (✓) item with the completed PQF:

_____ EMR documentation from your insurance carrier

_____ Insurance Certificate(s)

_____ OSHA 200 Logs (Past 3 Years)

_____ Safety & Health Program

_____ Safety & Health Incentive Program

_____ Substance Abuse Program

_____ Hazard Communication Program

_____ Respiratory Protection Program

_____ Housekeeping Policy

_____ Accident/Incident Investigation Procedure

_____ Unsafe Condition Reporting Procedure

_____ Safety & Health Inspection Form

_____ Safety & Health Audit Procedure or Form

_____ Safety & Health Orientation (Outline)

_____ Safety & Health Training Program (Outline)

_____ Example of Employee Safety & Health Training Records

_____ Safety & Health Training Schedule (Sample)

_____ Safety & Health Training for Supervisors (Outline)

Note: Owner checks items to be provided with PQF.

PQF EVALUATION
OWNER USE ONLY

DO NOT FILL OUT - OWNER USE ONLY

Contractor is:

☐ Acceptable for Approved Contractor List

☐ Conditionally Acceptable for Approved Contractor List

Conditions:

____

____

____

Reviewer: ____________________ Date: ____________________